Armours with special properties obtained by cold spraying on 52100 steel

Fabian Cezar Lupu
Corneliu Munteanu
Bogdan Istrate

Title: Armours with special properties obtained by cold spraying on 52100 steel

ISBN: 979-8-89248-581-4

Author:Fabian Cezar Lupu, Corneliu Munteanu,Bogdan Istrate

Cover image: www,pixabay.com

Publisher: Generis Publishing
Online orders: www.generis-publishing.com
Contact email: info@generis-publishing.com

Preface

This work entitled "Armours with special properties obtained by cold spraying on 52100 steel", is based on the Research Contract "Program in Materials, Additive and Secure Manufacturing, and Multiscale Materials Engineering (3ME)", Grant Number: W911NF2020024, a project carried out within the Research Laboratory of the U.S. Army at Northeastern University in Boston, in collaboration with several universities and research institutes in the United States of America (MIT, MMT, CTC, UCONN, VRC Metal Systems), as an efficient method for restoring and improving the material properties of armor for combat vehicles, through microstructural, mechanical and corrosion testing of the deposits made on 52100 steel.

The paper addresses the current stage of research, methods and equipment used in the research and experimental results. The introductory part includes the justification of the chosen theme and its purpose, so that the benefits of surface deposits in the military sector can be better understood.

The defence and maintenance industry of combat vehicles represents a vital strategic sector for national security and geopolitical stability. This industry focuses on the development, production, and maintenance of vehicles in the military sector. Combat vehicles, such as tanks, armored personnel carriers, and reconnaissance vehicles play an essential role in the capabilities of the armed forces, providing a mobile and powerful platform for military operations. This sector is functional and efficient due to the involvement and close collaboration between engineers, researchers and manufacturers, aiming to develop the most advanced technologies for military equipment. The constant maintenance and modernization of these vehicles are essential to maintain optimal performance during missions and to address technological advancements and increasingly complex threats. The defense and maintenance industry of combat vehicles not only contributes to a country's military strength, but also stimulates technological innovation, bringing significant benefits to the civilian sector.

Deposition methods show a trend of integration in the military and defense industry sector, with researchers and engineers continuously exploring innovative ways to improve the performance of armor materials. Cold Spray deposition technology represents a fascinating and efficient approach in this regard.

The Cold Spray deposition method, as a thermal spray process, is based on a dynamic increase in gas acceleration reaching supersonic speeds, leading to the generation of high kinetic energies that allow the consolidation of particles (raw material) into the base material, through a phenomenon of mechanical interlocking. This method becomes unique in that it is the only deposition technique that manages to deposit particles below their melting point, thus deposition temperatures are extremely low compared to other deposition techniques where deposition is a result of particle melting.

This technique becomes attractive for the military industry, as it can significantly contribute to improving the properties of armor materials, enhancing the level of protection against ballistic or explosive threats, so that combat vehicles benefit from a superior level of resistance and protection. Research conducted has led to remarkable results on the performance of armor materials, achieving efficient deposition of layers without damaging or compromising the base material. Furthermore, this technique allows for both improving the properties of the base material in manufacturing processes and excellent performance in maintenance processes regarding the restoration of various components with a high degree of wear.

The purpose of this study is to demonstrate the extent to which the Cold Spray deposition method is an efficient method for obtaining, the restoration and improvement of the armor material properties of combat vehicles through mechanical, microstructural and corrosion resistance testing of the deposits made on these materials.

Authors

CONTENT

CHAPTER 1 General considerations on the materials used for armour

The main weapon of protection for combat vehicles is the armour, so the armour defines the protection offered by them and is, usually, the key to the crew's survival, but also affects the dimensions, shape and, especially, the weight of the combat vehicle, all of which can affect the mobility and combat capability of the vehicle. In this context, a balance between the thickness/protection of the armour and the rest of the vehicle's performance is required. The evolution of armour is closely linked to the roles performed by combat vehicles, as well as the size and type of projectiles used by weapons intended to combat armoured vehicles. In Figure 1.1, a plate of armour perforated by projectiles can be observed [1,2].

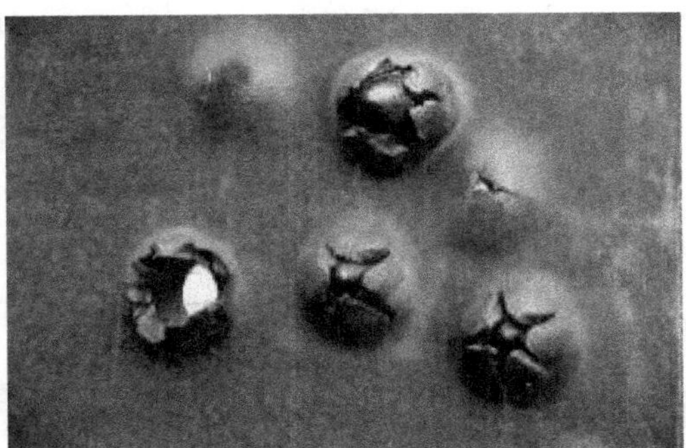

Figure 1.1 Penetrated armour plate [1]

Armours must respect a ratio between resistance and weight in order to maintain good manoeuvrability, thus, the thickest part is the frontal one where vulnerability is high, leaving other areas, such as the armour of the lateral parts of the chassis and turret thinner in order not to excessively increase in weight, while also being less protected. In Figure 1.2, armour elements on the Leopard 2 tank can be seen [1,3,4].

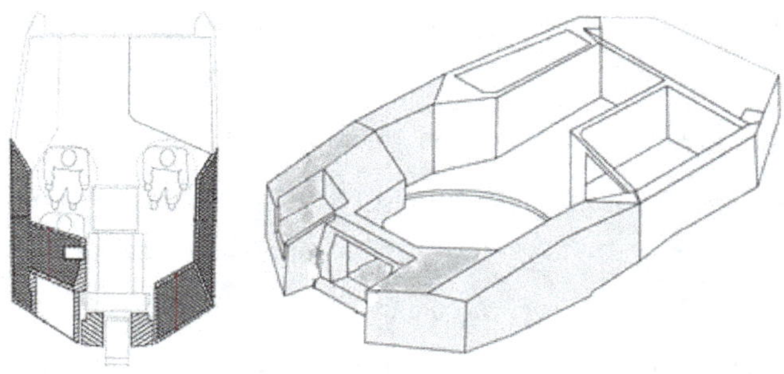

Figure 1.2 Leopard 2 armour [2]

1.1 Classification

Main types of armour:

Spaced armour has been used since the First World War, but it became famous on both tanks and German assault guns in the Second World War, the main quality being the disintegration/deflection of the projectile before reaching the main armour with minimal weight increase, while applique armour is used starting from the second conflagration, to supplement the main armour or add a layer with a different composition from the main one. Figure 1.3 shows the casing of the Abrams tank [1].

Figure 1.3 Abrams casing - machining [1]

Stratified/spaced armour protects against anti-tank missile attacks, the interior spaces leading to the deformation of the warhead before detonation or damaging the arming mechanism, preventing detonation. However, this armour is exposed to tandem warhead attacks, RPG-27 and 29 type (rocket-propelled grenade), which can penetrate multiple layers of armour, layers that can be observed in Figure 1.4 [1,5].

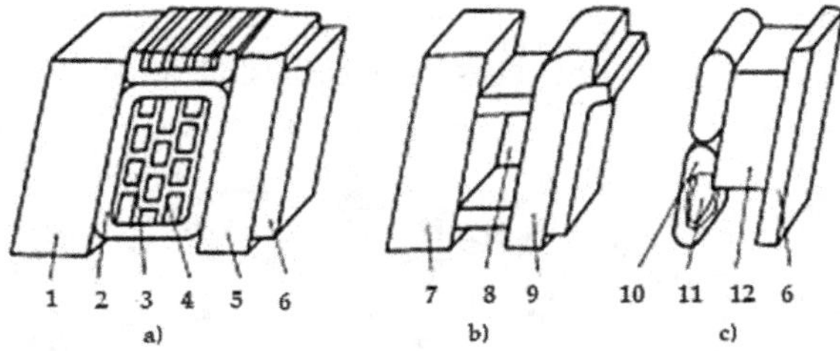

Figure 1.4 Stratified armour [1]

1 – base plate;

2 – aluminum or plastic casing;

3 – ceramic elements;

4 – adhesive;

5 – additional board;

6 – lining;

7 – additional obstacles (armour plate);

8 – cavity or polymer additions;

9 – front surface of the armoured box;

10 – metal cover;

11 – explosive charge;

12 – additional elements of protection and support.

The electrostatically charged armour operates based on two layers of electrostatically charged armour separated by an insulating layer. Upon penetration by a HEAT projectile of the strongly charged upper layer and the insulator, a powerful electric discharge is also produced, which discharges through a jet, significantly affecting it. The technology is still in its early stages, in terms of improvement. A form of spaced electromagnetic armour has been developed and tested by United Defense firm in the USA on the M2/M3 Bradley armoured vehicle, with a key aspect being that this armour, as can be seen in Figure 1.5, can withstand multiple hits and is also rechargeable. The armour is designed specifically for small caliber hits, up to anti-tank grenade hits, but further developments within the ARL program may also recommend it for protection against large caliber HEAT projectiles. [1,6].

Figure 1.5 T-80 turret without armour [1]

Reactive armour represents a solution with a partial degree of effectiveness, because it is characterised by a defense method composed of multiple tiles (elements) that are made up of an explosive placed between two plates, with the role of counterattacking the penetrating jet exerted by projectiles. This type of armour has the disadvantage that, in the event of an impact from a projectile, it emits a shower of fragments that are extremely dangerous for infantry, thus reducing the possibility of defending the tank by other soldiers when it moves at relatively low speeds and through easily vulnerable areas by enemies. Reactive armour is depicted in Figure 1.6 [1,7].

Figure 1.6 Reactive armour on the T-72 [1]

Composite armour, unlike other types of steel armour, has a slightly larger volume, it is made of several different types of materials, such as metal, plastic, ceramic, air compartments, and is designed to prevent penetration of projectiles, such as explosive-anti-tank projectiles, and has the advantage of being lighter than other types of armour. For example, on the Soviet T-64 tank, as it can be seen in Figure 1.7, a steel sandwich with a glass-hardened plastic core is laid out, which is made by pressure die-casting [1,8,9].

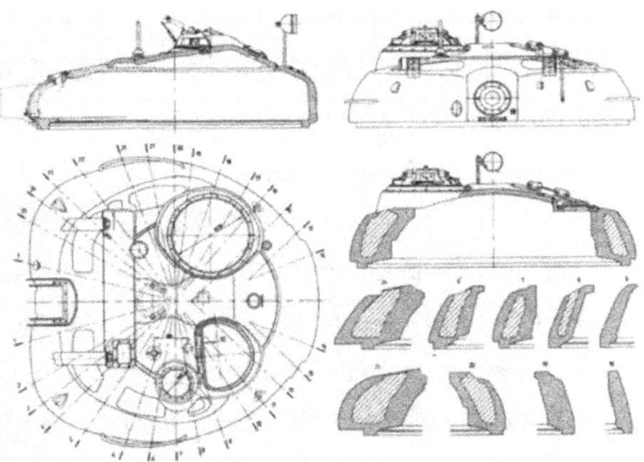

Figure 1.7 Armour plating on T-64 tank turret [1]

1.2 Materials used for armour

Even today, laminated steel armour represents, in most cases, the basic armour of the vehicle body to which other types of armor are added, sometimes in a sandwich structure [2,3].

Homogeneous armours are of two types, as follows:

- Cast armours;

- Laminated armours.

The first types of armours used were homogeneous, requiring a balanced ratio between toughness (to absorb the kinetic energy exerted by the projectile), hardness (to be able to break the projectile or its hard elements), and good fracture behaviour with a ductile character to facilitate the case of penetration by minimising the risk of forming splinters or fracture fragments [3].

Titanium is considered the metal with the best strength-to-weight ratio for armour, but due to its high price, it is used, almost exclusively, in the aircraft industry, especially for the A-10 Thunderbolt II and Suhoi Su-25 assault aircraft, in the form of a cylinder including the pilot's airframe, or as in the case of the MI-24 Hind attack helicopter [1,2,10,11].

Uranium, because of its high density, since it can absorb and dissipate the impact of projectiles, is used in the frontal armour of M1A1HA and M1A2 Abrams (DU) tanks, and it is also worth mentioning that uranium is also used for armour-piercing projectiles [1,2,12,13].

Aluminium has also been used in various combinations, in vehicles and light tanks, the last being foam, but although mechanically strong and light, it is less resistant to fire or certain types of armour-piercing projectiles [1,2].
Even **plastic** has been used in armour, laid over steel, and can stop armour-piercing projectiles by the hardness given by the granite composition, deflecting the projectile and slowing it down before it reaches the steel layer [1,2,5].

AMAP (Advanced Modular Armour Protection) is a modular composite armour made of nano-ceramics and steel alloys, 4th generation successor to Mexas and also developed by the German company IBD Deisenroth Engineering. AMAP contains high-hardness steel alloy (same level of protection, as ARMOX500Z High Hard Armour steel, at 30% reduced thickness), combined with a titanium-aluminum alloy, MAT 7720 (38% of the weight of an equivalent RHA laminated homogeneous armour), while titanium accounts for 58% of the weight of RHA laminated homogeneous armour, with the same level of protection [1,14].

ARMOX steel plates, also known as ballistic plates, are produced by the Swedish company on the Baltic coast, SSAB, advanced in the heavy plate branch [14,15].

1.3 Influence of alloying elements in steels used for armour

The alloying elements, depending on their quantities, give the desired properties to the steels. In Table 1.1, chemical compositions of some types of armouring are given and in Table 1.2, mechanical characteristics are shown.

Table 1.1 Comparative data on the chemical composition of some steel and aluminium armours [16]

Alloy steel IT-80	Al alloy		
	AA-5083	AA - 7039	Alcas GB – E 71 S
Ni 0.76 %	Mg 4.0 4.9 %	Zn 3.5 – 4.5 %	Zn 4.3 %
Cr 0.68 %	Mn 0.3 – 1.0%	Mg 2.3 – 3.3 %	Mg 2.5 %
Mo 0.34 %	Cr 0.05 – 0.25 %	Mn 0.1 – 0.4 %	Mn 0.3 %
	% Al	Cr 0.15 – 0.25 %	Cr 0.1 %
		% Al	% Al

Table 1.2 Main mechanical characteristics of some armourings [16]

Size feature	U/M	Al alloy		Steel IT-80	Cast steel B5-1456
		AA-5083	AA-7039		
Breaking strength	Kgf/mm^2	31	40	126.5	78.7
Yield strength	Kgf/mm^2	24.6	33.7	110.4	47.1
Specific shear strength	Kgf/mm^2	18.3	23.9	94.9	59
Longitudinal modulus of elasticity	Kgf/mm^2	$7,03.10^3$	$7,24.10^3$	21.10^3	21.10^3
Transverse modulus of elasticity	Kgf/mm^2	$2,7.10^3$	$2,69.10^3$	9.10^3	9.10^3
Density	Kgf/dm^3	2,657	2,74	7,83	7,79

ARMOX ballistic plates are produced by metallurgy, based on iron ore, and by means of blast furnaces, deoxidising steel in a LD converter and vacuum treated, results in a very clean steel. Through rolling mills, based on a complex technology, a fine-grained microstructure is obtained and, finally, in order to obtain the desired hardness/strength properties, the steel is subjected to heat treatments [14].

ARMOX steels are characterised by high strength in relation to hardness and are divided into the following types: ARMOX 370 T Class 1, ARMOX 370 T Class 2, ARMOX 440 T, ARMOX 500 T, ARMOX 520 T, ARMOX 560 T, ARMOX 600 T, ARMOX 620 T and ARMOX Advance.

ARMOX plates can be found in a range of hardness, from rolled homogeneous steel, with a hardness of 280 BHN, to UHH steel with a hardness of over 640 BHN and are mostly used in conjunction with materials, such as steel, aluminium, composite materials, etc. [14]. Table 1.3 shows chemical compositions and Table 1.4 mechanical properties for some types of ballistic plates complying with MIL-DTL - 46100E specifications. [17].

Table 1.3 Chemical composition of ARMOX ballistic plates [15]

Name	C max (%)	Si (%)	Mn max (%)	P max (%)	S max (%)	Cr max (%)	Ni max (%)	Mo max (%)	B max (%)
ARMOX 370T Class 1	0.32	0.40	1.20	0.010	0.003	1.0	1.80	0.70	0.005
ARMOX 370T Class 2	0.32	0.40	1.20	0.010	0.003	1.0	1.80	0.70	0.005
ARMOX 440T	0.21	0.50	1.20	0.010	0.003	1.0	2.50	0.70	0.005
ARMOX 500T	0.32	0.40	1.20	0.010	0.003	1.0	1.80	0.70	0.005
ARMOX 520T	0.32	0.40	1.20	0.010	0.003	1.0	1.80	0.70	0.005
ARMOX 560T	0.37	0.70	1.00	0.010	0.003	1.0	1.80	0.50	0.005
ARMOX 600T	0.47	0.70	1.0	0.010	0.003	1.5	3.0	0.70	0.005
ARMOX 620T	0.46	0.70	1.0	0.010	0.003	1.0	2.5	0.60	0.005
ARMOX ADVANCE	0.47	0.70	1.0	0.010	0.003	1.5	3.0	0.7	0.005

Table 1.4 Mechanical properties for some types of ARMOX ballistic plates [14]

Name	Hardness (BHN)	0.2 % Yield strength (N/mm^2)	Breaking strength (N/mm^2)	Extension
ARMOX 500T	480-540	min 1250	1450-1750	Min 8
ARMOX 600T	570-640	1500a	2000[a]	7[a]
ARMOX ADVANCE	HRC58-63	1600[a]	2250[a]	9[a]

Poplawski and colleagues conducted research on the fracture properties of Armox 500T steel regarding perforation issues, through an experimental-numerical procedure to develop the fracture surface in the triaxiality and Lode parameter space. The tests performed (including FEM tests, in addition to practical ones) on various types of samples, using three types of punches (sharp, flat, hemispherical), perforating 1 mm plates, show a sensitivity regarding the Lode parameter, so that the fracture properties depend on this parameter. The highest force required for perforation was exerted on the hemispherical punch, followed by the flat one and then the sharp one. Through a Kedzierski model, the highest force obtained was identified to be between the fracture models results of Skoglund and Iqbal. The perforated plates are shown in Figura 1.8 [18].

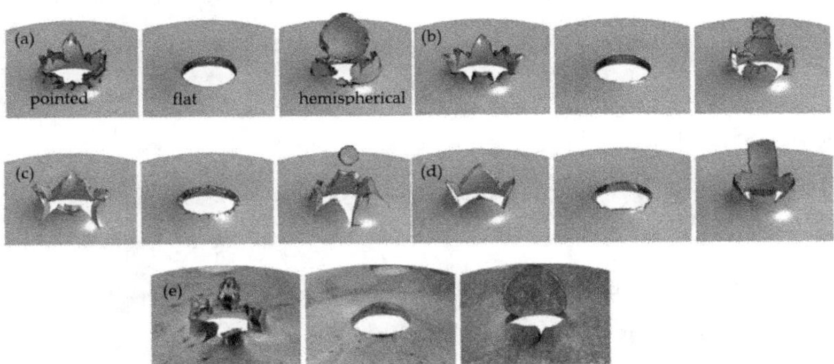

Figure 1.8 Skoglund and Iqbal models for plate perforation [18]

In the following graphs [18], are represented with grey solid line, the first trial of the experiment, with dotted line, the second trial of the experiment, with double dotted line, the third trial of the experiment, with yellow solid line, the FEM analysis of this study, with blue solid line, the Skoglund model FEM analysis, with green solid line, the Iqbal model FEM analysis and with red solid line, the Kedzierski model FEM analysis. The measured forces for punching plates are shown in Figure 1.9.

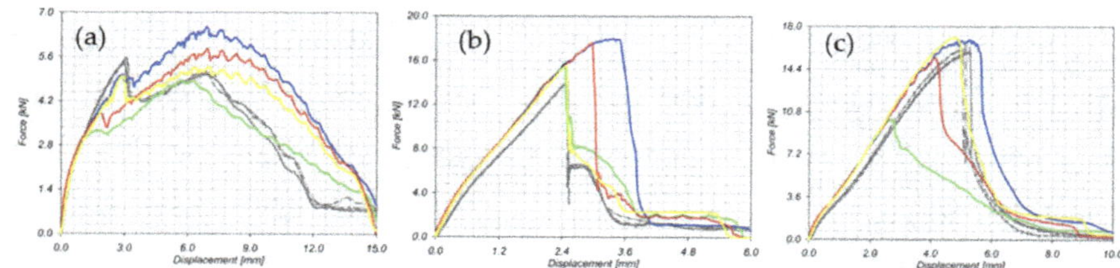

Figure 1.9 Forces measured for perforating plates:
(a) sharp punch, (b) flat punch and (c) hemispherical punch [18]

1.4 Improvement of armour properties by heat treatment

As noted in the previous sub-chapter, in order to achieve good impact characteristics on armour, there are a multitude of armour types, most of which are homogeneous, rolled steel plates, and in order to harden them, from what is available (literature) on a better understanding of the hardening properties of steels by heat treatment, it appears that rolled steel plates for armouring are subjected to the heat treatment of quenching and tempering, which leads to hardnesses of 28-35 HRC [19].

Quenching is achieved by heating hypoeutectoid steel to Ac_3+30-50°C and hypoeutectoid steel to Ac_1+50-70°C, followed by holding for austenitizing and high-speed cooling, which is above the upper critical speed ($v > v_{cs}$). The heating ranges are shown in Figure 1.10 [20,21].

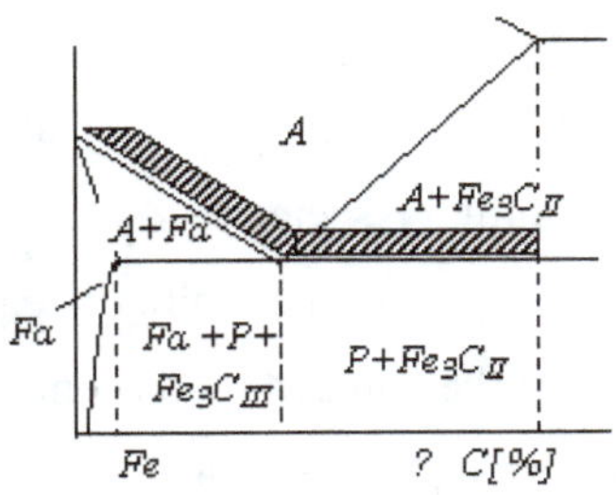

Figure 1.10 Heating ranges for the heat treatment of hardening [21]

Kruglijakow AA and co-workers carried out a study on the phenomenon of hot quenching of austenitic transformed steel, in order to observe the mechanical behaviour, but also the evolution of the RATE steel structure during deformation. After austenitizing at a temperature of 1150°, with holding and deformation at a temperature of 450°C, followed by slow heating to 750°C, with holding, deformation and, then, cooling to chamber temperature, zones of bainite, with a high density, can be observed in the steel structure. After cooling to room temperature, residual austenite does not exceed 5%. For RATE steel, these factors lead to an improvement of the hot hardening condition. The microstructure of the steel can be seen in Figure 1.11 [22].

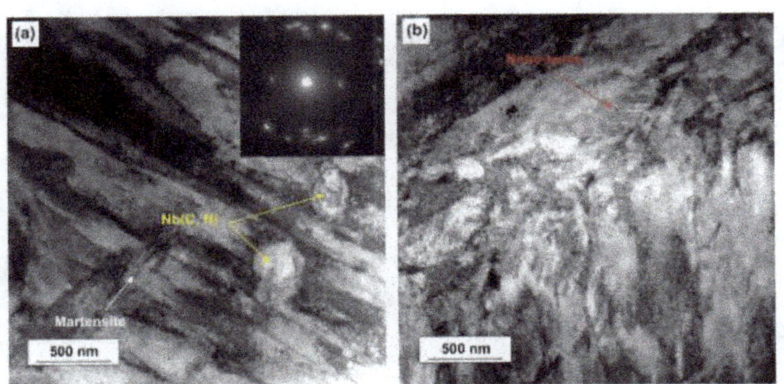

Figure 1.11 Image of the microstructure of RATE steel after austenitising at 1150 °C, followed by holding and deformation at 450°C and 750°C and, then, cooling to room temperature [22]

1.5 Partial conclusions

1. The information made available by specialised literature shows that there is a tendency of the military sector to opt as much as possible for the reconditioning of the components of the military equipment which involves substantial costs in the case of the complete replacement of the components, and the superficial deposits create the possibility to solve a part of these problems.

2. This research has the potential to contribute to scientific progress regarding the extent to which the Cold Spray method can improve armour elements of military equipment.

3. It was found that there were several types of materials used in the manufacture of armor and, at the same time, the specialized literature highlights, through extensive and various research, the superior properties that these materials have. From the review of the material types, it seems that ARMOX would be one of the best performing materials used in the manufacture of armours, reaching hardness values of up to 63 HRC.

CAPITOLUL 2 Modern spray technologies used to improve physico-chemical-mechanical properties

Thermal spraying is a group of coating processes for the purpose of producing coatings using as raw material powders, both metallic and non-metallic, which can be deposited in both melted and semi-melted state and have different properties, depending on the field of use. These surface deposition processes fall into three categories, as follows: cold methods, electrical methods and combustion methods [23].

In order to achieve thermal deposition, firstly, consumable materials are required and, secondly, an energy source is needed to achieve a temperature rise. In order to achieve the introduction of the material (powders) into the gun, and to obtain sufficient thermal energy to melt the material to be sprayed, gas is used, and in some cases, together with air. The gas, with its high velocity, sprays the powders in a melted state, in the form of fine droplets onto the surface used, and they solidify and adhere to it. The deposited layer has a lamellar structure, all the layers being tightly bound together [24,25].

Wear resistance, electrical resistivity, insulation at high temperatures, etc., are parameters that characterise thermal spraying, which are influenced by the method used in thermal spraying and, also, by the material used, while at the substrate level, during spraying (jet or flame), this process can bring changes to the material, such as: rapid solidification, material reduction, oxidation, etc. [26].

Thermal spray technology is a versatile technology, with process temperatures that span a large scale and provide a wide range of coatings. According to the evolution of this thermal spraying technology based on the type of energy used, 3 main groups can be distinguished, as follows in Table 2.1 [27].

Table 2.1 Thermal spraying technology and its acronyms [27]

Thermal spray technologies		
Cold methods	**Electrical methods**	**Combustion methods**
CGSM (cold gas spray)	APS (atmospheric pressure plasma jet coating)	flame spraying
HVAF (high velocity air fuel)	VPS (plasma spraying in a vacuum atmosphere)	HVOF (high velocity oxygen fuel)
WS (warm spraying)	LPPS (plasma spraying in low pressure atmosphere)	D-GUN (the detonation gun method)
CS (Cold Spray)		Arc spray

2.1 Cold Spray method

At the end of the 1980s, the technology of cold spray coating was developed by Allkimov and his collaborators. Through their research using metallic particles, it was concluded that the particles, instead of eroding the surface, adhered to it resulting in its coating [28].

Advanced by A.P. Allkimov, the patenting of this cold spray thermal technology took place in the USA, in 1994, and a year later, it was also patented in Europe [29,30].

Through this cold spray coating technique, the material jet is used in such a way that it does not result from the burning or ionization of the gas used in the process. The gas used or a mixture of gases that can be used in cold spray are compressed at a pressure of up to 3.5 MPa and at a heating temperature of up to 700°C. With the help of a convergent-divergent Laval nozzle system into which hot gas is introduced, it expands and reaches supersonic speeds. Powders are introduced into this nozzle system, after which the accelerated particles are deposited at values below their melting point [27].

During the coating process by the impact of particles with the surface undergoing spraying, they are subject to plastic deformation and the speed of these "unmelted" particles reaches values of 500-1500 m/s [31].

Cold spray (CS), as a thermal spray process, is based on a dynamic increase in gas acceleration reaching supersonic speeds and leading to the generation of high kinetic energies. The spraying device along with its components are shown in Figure 2.1 [32,33,34].

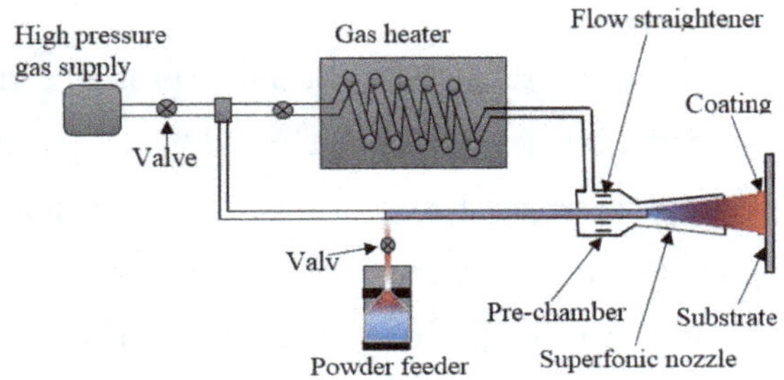

Figure 2.1 Schematic view of the Cold Spray device [25]

The main parameters of this spraying system (in an open environment) are the spraying distance which can have values in the range of 10-50 [mm] and the size of the powder which can have values in the range of 5-20 μm, these particles need to be characterised by plastic deformation properties at the moment of impact and must also have a low melting point [27].

The spray area achieved by the cold spray technique, reaches app. 20-60 mm2, with a productivity of 3-6 kg/h. This technology succeeds in spraying different materials such as alloys, ductile materials and also combinations of ductile and strong (or ceramic) materials that can be sprayed by this method, achieving an 80-90% yield. [35].

As a peculiarity in relation to ceramic materials is the installation that presents different characteristics, such as spraying pressure that includes values of less than 1 MPa (usually, 0.5MPa) and a flow rate that has lower values of about 0.4m^3/min [47].

Raw material powder and high gas consumption rate are the two main cost factors in cold spray technology. Helium has the advantages of efficiency and high quality microstructure, but it is also expensive. However, nitrogen, the most commonly researched cold spray propellant, is quite affordable [36].

When using the cold spray technique, due to its many benefits, a variety of industrial items can be produced and repaired, including turbine blades, pistons, cylinders, bearing components and shafts, etc. Various coatings can contribute to characteristics including hardness, wear and corrosion resistance and thermal conductivity [7].

For example, various tests show that cold spraying of W is quite ambitious due to the hardness and brittleness of the material at very high temperatures. In contrast, cold spraying can create thick and very dense tantalum (Ta) coatings, which are already used in the space, chemical and nuclear industries, as well as in medical applications. Metallographic cross-sections of coatings created by separate feeding of coarse W and Ta powders, where coatings as thick as 2 mm were produced, are shown in the microscopic images in Figure 2.2. All coatings show low porosity and an increase in volume fraction of W, with increasing percentage of W powder, as can be seen from the micrographs [37,38].

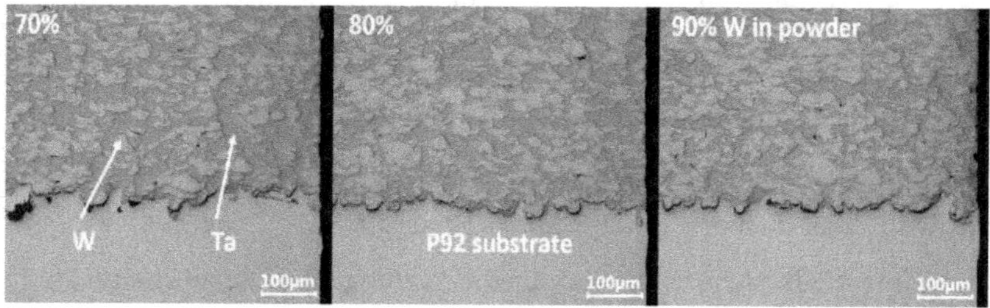

Figure 2.2 Cold-spray coatings with varying amounts of tungsten and tantalum, seen in cross-section [37]

In research by Agiwal et al. repairing cracks in a 304 L austenitic stainless steel, two methods were investigated: cold spray deposition (CS) and surface friction (FS). Both methods are solid-state, and the low heat input avoids the negative impacts of solidification, such as compositionally segregated microstructures and shrinkage stresses. Cross-sectional images of the coating-substrate contacts fixed by the two procedures are shown in Figure 2.3. In the steady-state region, the consumable rod friction surface generated coatings that were 325 µm thick and 2.3 mm wide. The absence of porosity in the coatings and the smooth interface between the coating and the substrate indicate a strong connection. Figure 2.3 (c) demonstrates how the FS procedure can repair a dummy crack. At a depth of 75 μm below the coating-substrate contact, the fracture point was seen to tilt. High residual compressive stresses at the surface, caused by surface friction, are believed to have contributed to the plastic deformation of the substrate and the bending and closing of the fracture. A high-density cold spray coating (porosity level less than 0.3%) covering the transverse crack, can be seen in Figure 2.3 (b). The coating had a thickness of ~620 μm. As it approached the coating-substrate contact, the crack width decreased. At a distance of about ~90 μm above the coating/substrate contact, the crack eventually disappeared. As shown in Figure 2.3 (d), a few severely damaged powder particles were found inside the crack. Two potential processes for crack repair include compressive stress from the CS process and deformation of the substrate upon opening the crack, followed by application of a stainless steel coating. For example, the substrate may become smaller as a result of the lateral displacement it achieves after the powder particles hit it at the crack opening. Due to the peening action, complete coverage can be achieved during particle impact once the crack opening size is smaller than the powder particle size (<44 μm). Phase preservation and lack of deformation-induced martensite distinguish cold spray coatings from friction surface coatings [39,40].

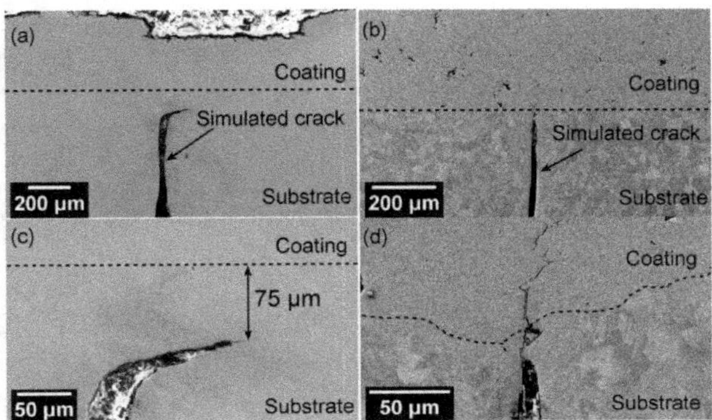

Figure 2.3 SEM cross-sectional images of areas that have been restored using cold spray deposition (b,d) and friction surfacing (a,c) [39]

Wenpeng and collaborators conducted a study due to the requirements of the spray gun to facilitate spray control over a long period of time, these requirements being found through studies that the spray injector (component that is part of the gun assembly) leads to a disturbance of the flow fields and particle impact parameters. After conducting the tests, it was concluded that the particle size, the injector spray orifice diameter and the pressure discrepancy between the gas supplying the powder and the main working gas led to a significant influence on the particle distribution both near the spray nozzle throat and on the coating. Figure 2.4 illustrates how the range of particle deposition on the substrate increases as the injector inner diameter decreases, in addition, as the injector inner diameter increases, so does the particle departure gradient from the nozzle axis upstream of the nozzle. As a result, the powder injector with a smaller inner diameter can, in the cold spraying process, not only increase the production accuracy of the parts, but also, to some extent, reduce the risk of nozzle blockage or sticking upstream of the nozzle [41,42].

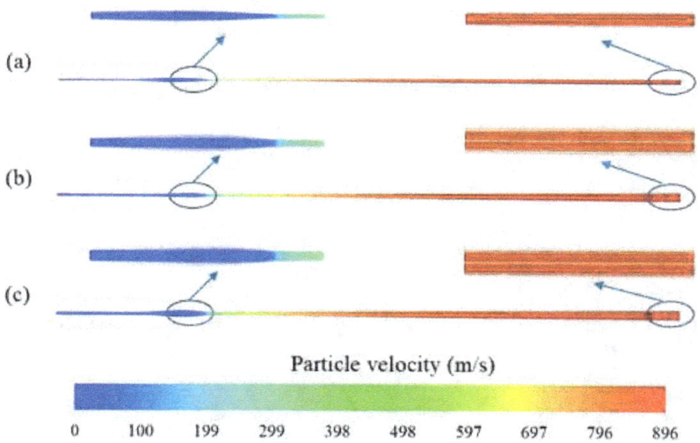

Figure 2.4 The particle jet at different inside diameters of the powder injector are coloured according to particle velocity: (a) 0.5 mm; (b) 1 mm and (c) 1.5 mm.
Particle sizes range from 5 to 60 µm [41]

Through these tests, it has been shown that improving particle impact velocity, as well as reducing the risk of nozzle clogging with a particle concentration close to the nozzle, is optimised by reducing the internal diameter of the powder injector and using a lower differential pressure and, at the same time, the use of particles with diameters between 20-40 μm also leads to improved accuracy [41].

An important aspect of this surface spraying technology is that it expands the range of applications in terms of thermal spraying, applying in industries such as automotive, aerospace, medical and petrochemical industries, helping both production and restoration of various components, such as machining defects, casting defects and even the possibility of restoring moulds [43].

Essentially, cold spraying is distinct from other spraying methods, because it operates at much lower temperatures than thermal spraying, primarily using kinetic energy to create the bonded coatings in a solid state, rather than melting the particles and, then, resolidifying them. Figure 2.5 shows the temperature and deposition rate values for different spray methods, so that a comparison can be made with the Cold Spray method. [44].

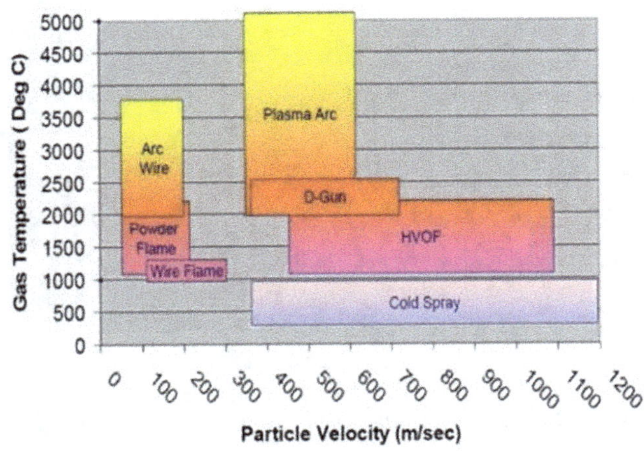

Figure 2.5 Particle temperature and velocity for different deposition methods

[44]

2.2 Flame spraying

Flame spraying technology, invented in 1910, by Dr. Schoop and collaborators, originated with the creation of a device for the thermal spraying of metal wires, a device based on the melting of a metal wire by means of a flame obtained by burning fuel with oxygen, from where the resulting melted metal was atomised by means of compressed gas propelling the droplets of melted metal into the surface layer of the surface on which the deposition is made. This flame spraying technology later gave rise to High Velocity Oxygen Fuel (HVOF) technology [23,45].

To achieve the melting of the material to be sprayed, this method relies on the heat that is obtained by burning the fuel mixture which can be: oxygen-acetylene, oxygen-hydrogen or oxygen-propane, the temperature reached during the spraying being approx. 3000°C. The particle velocity, which varies according to the density and geometry of the particles, is not very high, ranging from 40-100 m/s, these low values resulting in a high porosity of the layers and a high level of oxides on their microstructure. Figure 2.6 shows a section of the sputtering device and the wire that is pulled into the device [24,46,47].

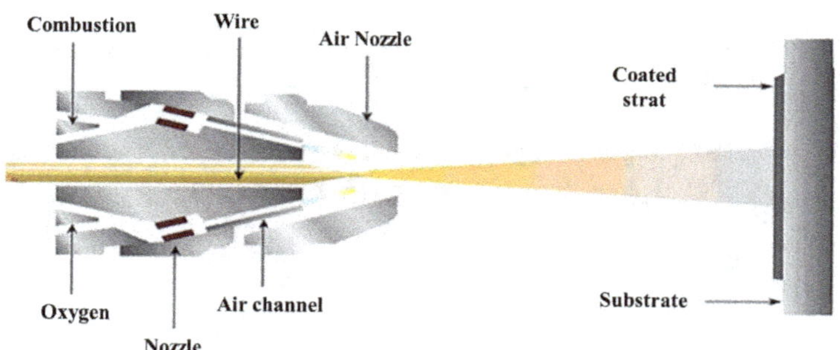

Figure 2.6 Flame spraying using wire as raw material [24]

Flame spraying using powders as raw material consists of a multi-hole device, whereby the mixture is preheated by acetylene and oxygen burning around the nozzle device, and the powder particles are injected axially, with the gaseous fuel. The spraying distance is 120-250 [mm] and the substrate temperature in an open atmosphere is up to 250°C. Figure 2.7 shows a section of the powder spraying device [24,27,47].

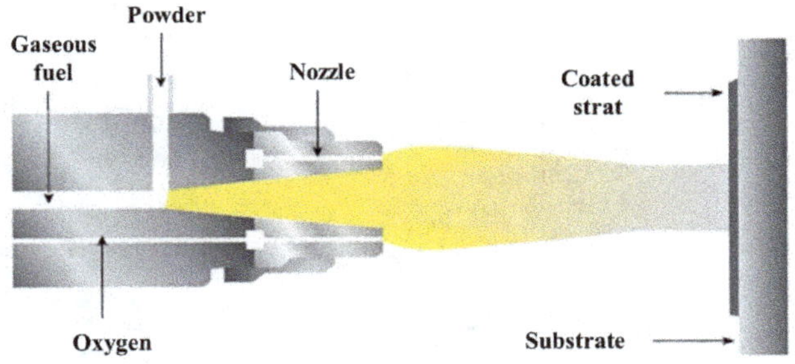

Figure 2.7 Flame spraying using powders as raw material [24]

2.3 High Velocity Oxy Fuel (HVOF)

Introduced in 1958, by Union Carbide, HVOF spraying did not become commercially available until the early 1980s, with Browing's Jet Kote system (1983), which ran on oxygen and gaseous fuel [47].

The HVOF spray technology, in order to obtain a combustion jet that manages to reach temperatures of 2500-3100 °C, uses gaseous fuels, such as hydrogen, propane, etc., which together with oxygen, lead to these temperature values. In order to obtain the supersonic gas jet from which a high particle velocity is generated, the device has exhaust ports with a diameter of approx. 8-9 mm, and both wire and powders can be used as raw material. Figure 2.8 shows types of guns used in HVOF thermal spraying, such as: (a) Browning Jet Kote, (b) axial powder injection and combustion chamber, (c) axial chamber with radial powder injection [23,48].

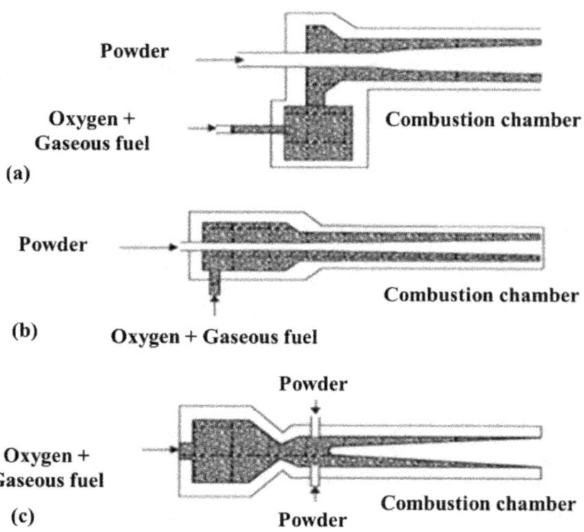

Figure 2.8. Types of guns used for thermal spraying by the HVOF method [47]

2.4 Detonation Gun (D-gun)

D-Gun sputtering technology, developed in Russia and introduced in 1950, by Gfeler and Baiker working for Union Carbide, has an exceptional benefit through the detonation process, where carbides applied to the surface, exhibit very good adhesion forces, hardness and density [24,47].

This spraying system involves powder capsules being placed in a 1 m long cylinder, together with oxygen and fuel gas, which is usually acetylene, and when the mixture is ignited by a spark that causes a controlled explosion, which results in the charge being ejected from the cylinder. Figure 2.9 shows a spray gun seen in section [23].

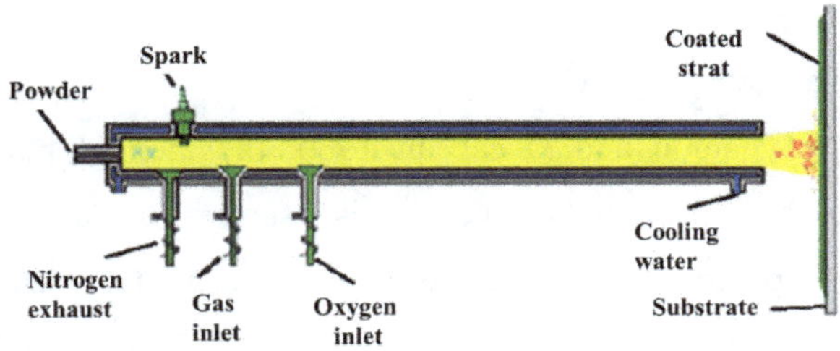

Figure 2.9 Gun used in the D-Gun spraying process [49]

Unlike the classic D-Gun which can produce 1-15 detonations/sec, modern devices reach performance values of up to 100 detonations/sec. These new systems can produce working flames at high velocities and generate a high spray rate. The main parameters are the spray distance, which has values of 100-300 mm, and the powder size ranging from 5-60 µm. [45,50].

2.5 Wire Arc Spray

This method, invented in 1910, by Schoop, uses two wires that are placed in an electric arc where later, with the help of a compressed gas (air or nitrogen), the material is sprayed onto the surface used. The wires (cathode and anode) are melted and broken into droplets by the gas jet, which can have a flow rate of 0.8 – 3 m3/min, and the resulting surface is quite porous. Figure 2.10 shows a gun used for the Wire Arc Spray process, seen in section [47,51].

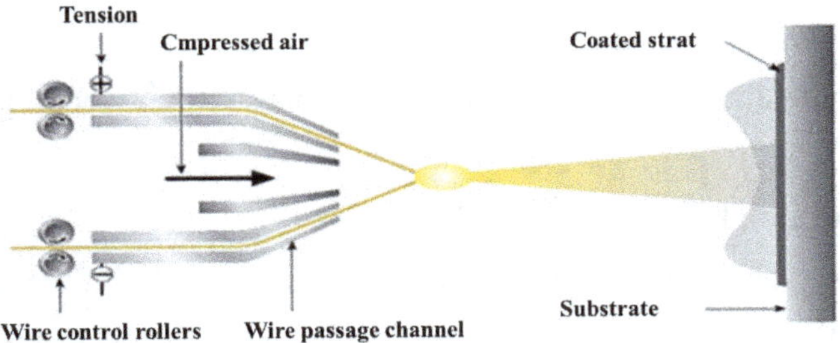

Figure 2.10 Gun used in Wire Arc Spray process [24]

2.6 Arc Plasma Spray (APS)

Invented by Giannin and Ducati in 1960, the thermal plasma spraying process is the most commonly used and is usually carried out in a normal atmosphere. The process forms an electric arc between a tungsten electrode and a copper anode, both cooled with water, and when the two electrodes are closed due to the electric arc where gas dissociation and 26pprox.26on occurs, it leads to the formation of plasma when they reach temperatures of 7000-8000°C. Plasma is a neutral combination (same number of electrons as ions) of molecules, ions, atoms, photons and electrons [23,45,47,52,53].

Thermal plasma spraying can use a wide range of raw materials and a particle distribution that can vary in size, because this spraying technology has a wide range in terms of changing plasma temperatures and particle velocities. The plasma spraying device is shown in Figure 2.11. [24].

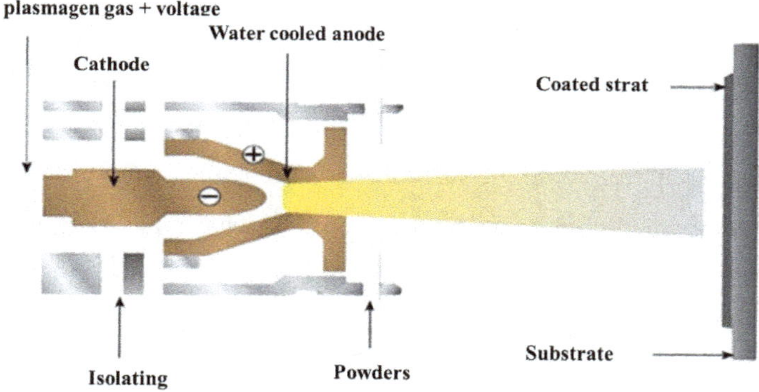

Figure 2.11 Device used in thermal plasma spraying process [24]

2.7 Vacuum Plasma Spray (VPS)

In 1974, Muehlberger of today's Sulzer Metco (formerly Electro Plasma Inc) developed the method of thermal plasma spraying in a controlled atmosphere. In the other method of plasma spraying, but which is carried out at atmospheric pressure, oxidation occurs, which is detrimental to the coatings and adhesion of alloys and metals to the substrate. [54,55].

The controlled-atmosphere plasma spraying device is located in a sealed chamber that is cooled with water and has an internal pressure of 10-70 Kpa. The plasma temperature reaches values of only a few thousand degrees Celsius, the spraying distance reaches values of 500 mm and the thickness of the sprayed layer reaches values between 20 – 90 µm. The thermal plasma spraying device in a controlled atmosphere is shown in Figure 2.12 [45,56].

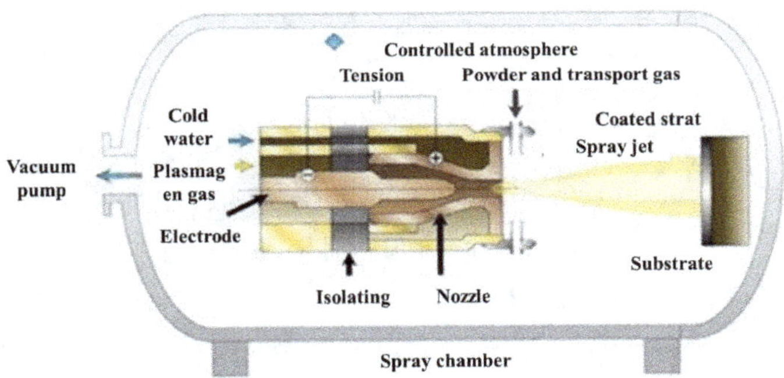

Figure 2.12 Controlled atmosphere plasma thermal spraying system [24]

2.8 Low Pressure Plasma Spray (LPPS)

The LPPS spraying process is a special coating technique intended for somewhat more special applications due to the fact that the density of the deposited layer must present characteristics similar to those of a forged alloy. By means of this technique, deposited layers are made with a very low porosity and oxidation, and the operating pressure, respectively 50-200 mbar, leads to obtaining the excellent quality of these layers [24].

Just as any technology also comes with a series of disadvantages, in the case of LPPS spraying technology, we can mention the fact that the difficulty, in terms of the execution cost of the work, cannot be neglected, it can reach values even 20 times higher compared to the APS spraying technology, at the same time also creating a difficulty in melting refractory materials, because in the vacuum environment, through the Knudsen effect, the heat transfer to the particles is limited. The scheme of this spraying technique is shown in Figure 2.13 [56].

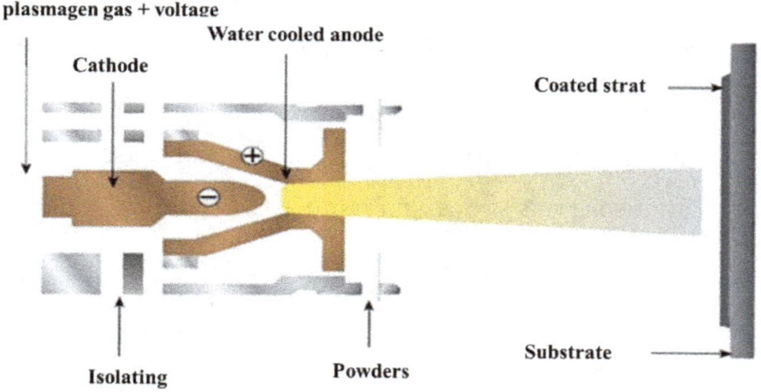

Figure 2.13 Scheme of LPPS thermal spray technology (scheme similar to APS method) [24]

2.9 High Velocity Air Fuel (HVAF)

The HVAF thermal spray method is a hot spray method, where the operating temperature range is between the cold spray technique and the HVOF technique. Through an air-fuel jet, axial injection powders are injected at temperatures of 1900-1950°C, used in HVAF guns. Similar to cold spraying, the HVAF system can apply non-oxidizing materials, efficiently applying carbide-based materials through the fact that the air-fuel jet produces a small amount of oxides as opposed to the high-temperature oxy-fuel jet [57].

Bandyopadhyay and Nylen created a very detailed model of the flame in terms of combustion diffusion [58].

Eddy's dissipation concept, in terms of creation and destruction rates, was considered by Bandyopadhyay and Nylen to be the calculation in which the dominant value of the two physical expressions, namely the dissipation rate (ε) and the kinetic energy of turbulence (K) is used [59].

For both HVOF and HVAF, it is assumed that most of the reactions take place in the combustion chamber, and the extent to which the reactions progress is a result of a chemical equilibrium model, an assumption that is exposed by the fact that, unlike the other components that are part of the device, in the combustion chamber, the gas stays for the longest time. The oxidation level of thermal spraying processes is shown in Figure 2.14 [60].

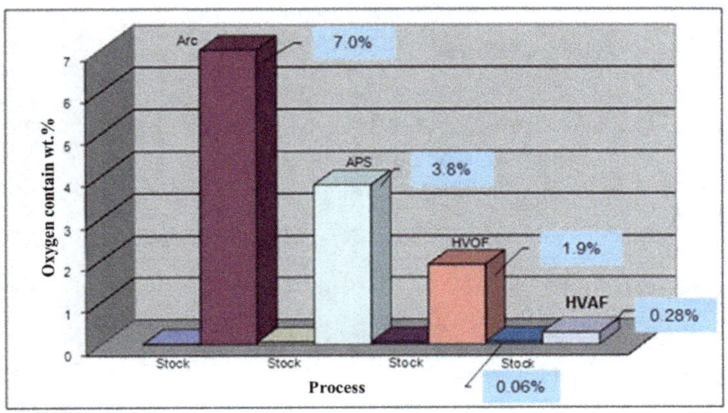

Figure 2.14 Graph showing the oxidation level of thermal spraying processes [57]

2.10 Warm Spraying (WS)

Through HVOF high-speed spraying technology, the hot surface spraying process was born. The characteristic of this process consists in controlling the temperature of the gas that propels the particles in a range between 726°C – 2026°C, control achieved through the HVOF system in which inert gas is injected into the gas jet in the combustion chamber and through this range of varying the temperature, is the advantage of the possibility of using several types of powders [61]. Figure 2.15 shows the operation mode of the hot spraying device (HVOF), in two phases, to which is attached the table with the expression of the mathematical model of the regions, a mathematical model that includes two parts, namely: the representation of the simulation in terms of the flow of gas-particles both inside and outside the spraying device, on the one hand, and on the other hand, depending on the flow range of the particles, of the calculation regarding the mass gain through the oxidation of the particles, characteristics that are obtained in the first section.

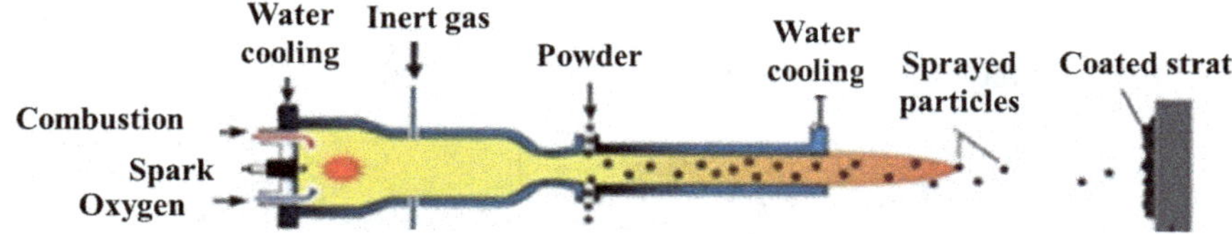

Figure 2.15 Hot Air Spray System (HVOF) with two-phase schematic and region mathematical model table [62]

The hot spraying method, through the wide range of deposition temperature values, removes the disadvantages between cold spraying and HVOF spraying technology [62]

2.11 Partial conclusions

1. Thermal deposition methods have the potential to improve and restore military components, all of which can solve numerous problems. However, due to the high temperatures during deposition, sometimes exceeding 10000°C, some applications in which they are used, are conditioned.

2. Thermal deposition methods manage to successfully deposit superficial layers that improve the properties of the base material, depending on the type of powders used.

3. Due to the thicknesses of the deposited layers where, for example, in the case of plasma jet spraying (APS), thicknesses of approximately 50 μm are reached, the excellent use is allowed to increase the material properties of the components that are subjected to severe regimes, but it becomes limiting in regarding the dimensional restorations of the components that require significant thicknesses.

4. The Cold Spray deposition method proves to be a fairly effective method which, at first sight, succeeds in complementing the other methods by the fact that it can create coverages of the order of tens of millimeters.

5. The Cold Spray method becomes unique in that it is the only method of depositing particles below their melting point, this having the lowest deposition temperatures, reaching a maximum of 31pprox.. 700°C. The principle underlying this method is the acceleration of the particles at supersonic speeds, so that they deform plastically and intertwine in the material, resulting in the creation of dense and closely connected layers.

CHAPTER 3 Types of materials used to improve properties by the Cold Spray surface deposition method

3.1 Properties of 52100 steel, as a base material for surface spraying

This type of steel, called ASTM A295 (52100), is alloyed with Chromium and has a high carbon content, which gives it high hardness and is characterised by high resistance to wear and fatigue, thus meeting standards such as ASTM and SAE. 52100 steel has a density of 7.81 g/cm3 and a melting point of 1424°C [63,64]. The exposure of the alloying elements that enter the composition of 52100 steel, as well as the equivalents together with their standards are specified in Table 3.1.

Table 3.1 The percentage of alloying elements included in the chemical composition of 52100 steel and its equivalents [64,65]

Standard	Type	Fe (%)	C (%)	Mn (%)	P (%)	S (%)	Si (%)	Ni (%)	Cr (%)	Cu (%)
AISI	52100	bal.	0.940	0.340	0.009	0.007	0.200		1.400	
AISI [66]	316L	bal.	0.022	1.583	0.037	0.0004	0.516	12.710	17.560	
ASTM A295 [67]	52100	bal.	0.93 - 1.05	0.25 - 0.45	0.025	0.015	0.15 - 0.35	0.25	1.35 - 1.60	0.30
DIN 17230 [68]	100Cr6/ 1.3505	bal.	0.90 - 1.05	0.25 - 0.45	0.030	0.025	0.15 - 0.35	0.30	1.35 - 1.65	0.30
IT G4805 [69]	SUJ2	bal.	0.95 - 1.10	0.50	0.025	0.025	0.15 - 0.35		1.30 - 1.60	
BS 970 [70]	535A99/ EN 31	bal.	0.95- 0.10	0.40 - 0.70			0.10 - 0.35		1.20 - 1.60	

A factor that contributes to highlighting the superior qualities of 52100 steel, characterized by mechanical properties in terms of elastic stiffness, is the modulus of elasticity that reaches high values, namely 210 GPa, making it a rigid material, responding well to machining processes, having also important characteristics on compressive strength, which, for example, unlike most aluminum alloys, is almost twice as strong, with a bulk modulus value of 160 GPa. These properties, both physical and mechanical, create performance that makes 52100 steel an ideal steel for highly demanding parts, such as aircraft bearings, ball screws, etc. [63,64,71].

In the literature, between the 1940s and 1990s, it was found that in bearing steels, during cyclic loading, due to rolling stresses, a strong contact fatigue stress (RCF) occurs, resulting in substantial microstructural changes [72]. Around the 1980s and 1990s, research conducted by numerous institutes on heat treatment concluded that fatigue performance is undoubtedly influenced by material condition, and some studies show that surface hardening leads to longer lifetime (RCF) to unlike the fully reinforced elements [73,74].

Viktorija et al. [72] conducted research on the microstructure of 52100 steel, including a joint heat treatment process with austenitizing at 830-860°C, followed by oil cooling at 60°C and holding for 2h at 170-220°C, after that, the samples were analyzed by scanning electron microscopy and it was observed that the steel has a microstructure composed of martensite and primary spheroidized carbides with homogeneous distribution (Fe, Cr)$_3$C, tempered carbides and 10-12% residual austenite. Through SEM, SE and BSE images, primary (Fe, Cr)$_3$C spheroidal carbides are exemplified by arrows, these primary iron-chromium spheroidal carbides being unaltered, as shown in Figure 3.1.

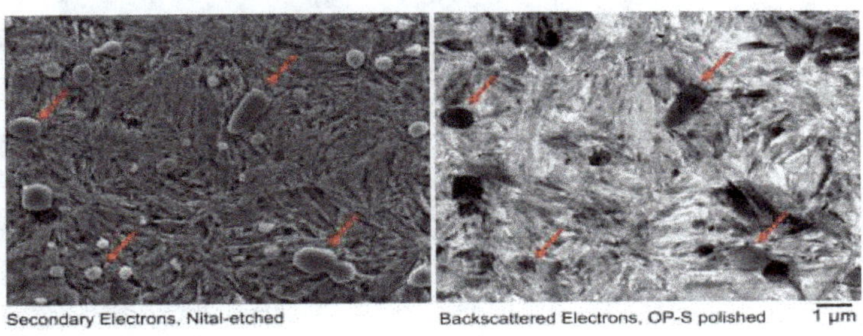

Figure 3.1 Images of the microstructure of 52100 steel [72]

Kiranbabu and collaborators approached for a better understanding, the mechanism of carbide decomposition in 52100 steel (100Cr6), a mechanism that is crucial for obtaining corrosion-resistant steels encountered in industrial applications. The decomposition of cementite is investigated through a modified thermal treatment in order to generate a soft, annealed microstructure, in which spherical cementite and lamellar cementite precipitate. In order to investigate the effects of composition, morphology and size, as well as the microstructure of the matrix in order to decompose cementite, the microstructure was subjected to a severe plastic deformation induced by high pressure torsion-HPT (P=7.56GPa, Ymax~50). After the investigations, it was concluded that, by adding Cr and Mg, it leads to an increase in the stability of cementite. Regarding the wear of the spherical cementite precipitates, no wear was observed, the pearlitic matrix presenting a hardness of 4.7 GPa (in view of the wear behavior, as a decisive role, is the hardness of the matrix around the precipitates). In Figure 3.2, the Y~50 lamella is investigated where (a) represents deformed spherical cementite, and (b) the precipitate of deformed spherical cementite in (a). A precipitate smaller than (~100nm) does not indicate any internal deformation [75,76].

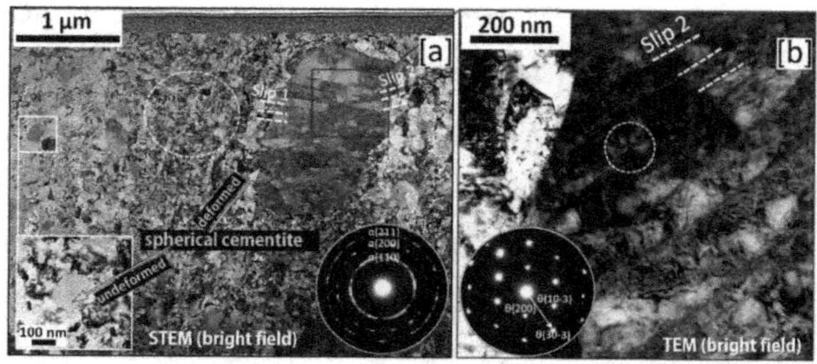

Figure 3.2 TEM image with diffraction analysis of Y~50 sample. Spherical cementite [75]

Solis-Romero J. and collaborators carried out a study in order to deposit hydrogenated carbon (similar to diamond) on AISI 52100 steel, and in order to achieve real working conditions, they used a "grey-fuzzy" hybrid method, and the process took place in an argon/acetylene atmosphere, by means of a Hauzer Flexicoat850 system, the coating being produced by means of plasma chemical vapor deposition (PECVD) technology for enhancement, C_2H_2 gas being prepared at 0.3 Pa. After the test, the successful deposition of a-C:H led to obtaining a layer of ~3μm with an elasticity modulus of up to 175 Gpa, with good adhesion properties under critical load, respectively 72.6 N (the industrial requirement being 50 N). The cross-section of the coating is shown in Figure 3.3 [77].

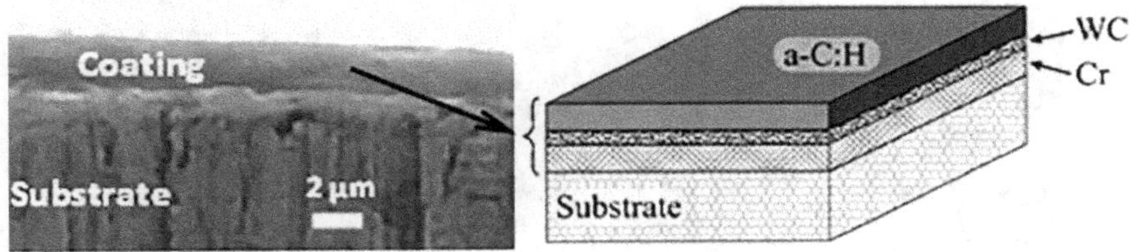

Figure 3.3 Cross-section a-C: H cover [77]

Rahbar et al. conducted research on a 52100 steel substrate, using WC-12%Co based powders that was thermally sprayed before friction stir processing (FSP) was performed on this layer. Wear resistance and hardness were tested before and after FSP. FSP combines the sputter layer with the substrate decreasing porosity and improving hardness and wear performance, according to optical and SEM analyses. Three different structures of the sputtered layer are shown in Figure 3.4 (a), the WC-12%Co layer is the upper porous layer, and for improved adhesion, the intermediate layer is a Ni-Al composite, and the lower layer is the base metal. According to Figure 3.4. (b), the refined zone containing martensite and WC particles extends to about 1.2 mm below the surface of the sputtered layer after FSP. WC is dispersed in a martensitic matrix in this image. When comparing the FSP structure with the sputtered structure, neither the layer boundaries nor the porosity are clearly defined. Because a totally dense microstructure was created by thermo-mechanical deformation, FSP solved two disadvantages of the spray coating, namely: porosity and poor adhesion. The WC particle size is shown in Figure 3.4 (c) and (d) before and after FSP, respectively. It is obvious that the FSP treatment can reduce the original WC particles to sub-micrometer size, and this will affect the hardness and wear performance. During FSP, smaller WC particles may also experience plastic flow in a Co binder at high temperatures, which would promote a full-thickness microstructure and improve WC particle spreading. This would provide high microhardness [78].

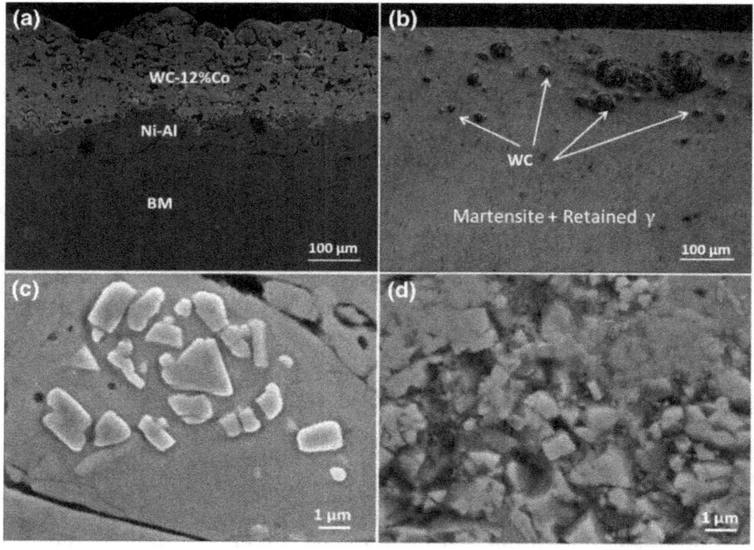

Figure 3.4 The structure of the sprayed layer (a) before and (b) after FSP and the WC particle size (c) before and (d) after FSP [78]

3.2 Powders used for coatings that improve mechanical properties

In order to improve the mechanical properties, different types of powders were developed such as WIP-C1, WIP-C2, WIP-W1, WIP-W2, WIP-BC1 (WIP-BC1 used as a bonding substrate), DARC-AL6061-G1H1, METCO 82VF-NS, METCO430NS, etc. below, detailing some of these.

In order to ensure a much closer connection between the base material and the coating, coatings are made with powders specially designed for creating this connection, such as **WIP-BC1** powders, based on chromium carbide and agglomerated with nickel, powders that present a high level of resistance to wear and impact, being specially designed for establishing this hard bonding layer between the base substrate and the coating layer in which the WIP-C1 and WIP-

C2 type powders are used [79]. The morphology of the powder can be seen in Figure 3.5, the chemical composition in Table 3.2 and the mechanical properties in Table 3.3.

WIP-BC1 has high fluidity and is intended exclusively for the production of coatings as a base material, it can be cold sprayed with nitrogen or helium [**Error! Bookmark not defined.**].

By means of this hard bonding substrate, the coating layers adhere very well and the risk of delamination of the deposited surface layer and interfacial porosity is excluded [**Error! Bookmark not defined.**].

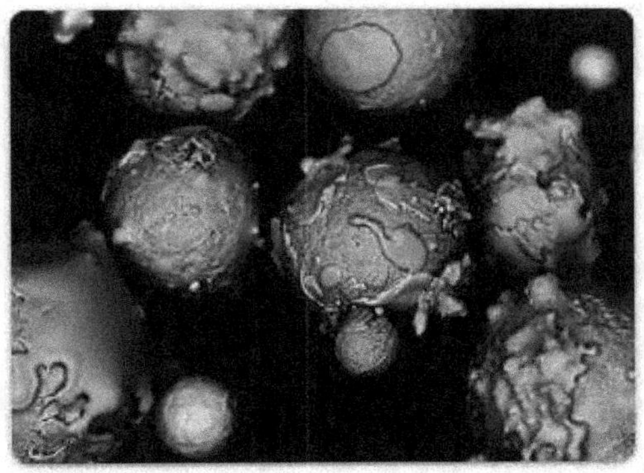

Figure 3.5 WIP-BC1 powder morphology [**Error! Bookmark not defined.**]

Table 3.2 WIP-BC1 powder morphology [**Error! Bookmark not defined.**]

Concentration of elements (%)	Cr (%)	C (%)	Ni (%)
WIP-BC1	59-76	2-5	22-35

Table 3.3 Mechanical properties of WIP-BC1 powders [**Error! Bookmark not defined.**]

Material	Geometry	Adhesion
Carbide	Spheric	>10 [ksi]

WIP-C1 powders, based on chromium carbide, present a high level of performance both in terms of wear resistance and impact resistance. The powders are intended for the cold spraying process that achieves a dense coating that can be sprayed with both helium and nitrogen, being able to obtain coatings with a thickness of approximately 6 mm, at the same time, when spraying is carried out on a hard layer, a bonding substrate using WIP-BC1 powders is required. This type of material, which is based on chromium carbide, is agglomerated with nickel in order to obtain hard properties, resistance to both wear and impact. The coatings made with this type of powders (WIP - C1) allow the subsequent processing of the surface which can be done both by milling and by turning (at the same time and/or grinding). The morphology of the powder can be seen in Figure 3.6, the chemical composition in Table 3.4 and the mechanical properties in Table 3.5 [**Error! Bookmark not defined.**].

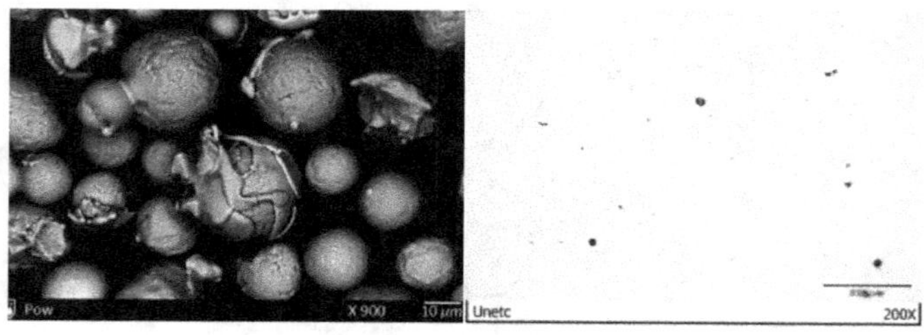

Figure 3.6 Morphology of WIP-C1 powders (a) and the sprayed surface (b)
[**Error! Bookmark not defined.**]

Table 3.4 Chemical composition of WIP-C1 powders [**Error! Bookmark not defined.**,80]

Element concentration (%)	Co (%)	Cr (%)	Fe (%)	Ni (%)	W (%)	Al (%)	B (%)	Mo (%)	Si (%)	V (%)	C (%)
WIP-C1	0.02	68.45	0.08	27.84	0.01	0.06	0.04	0.01	0.43	0.01	3.00

Table 3.5 Mechanical properties of WIP-C1 powders [**Error! Bookmark not defined.**]

Material	Geometry	Hardness	Porosity
Carbide	Spheric	375-425 [HV]	<1% (N2)

WIP-C2 powders, which are based on chromium carbide, show superior properties in terms of hardness, wear resistance and impact, they are agglomerated with nickel-chromium, powders intended for cold spraying, they can be sprayed both with helium as well as nitrogen and beforehand, when spraying on a hard layer, a bonding substrate using WIP-BC1 powders is required. In order to increase the hardness, while keeping the same density and low porosity, unlike the other type of powders mentioned before (WIP-C1), the chemical composition of the WIP-C2 powders has been optimized, especially the increase of the chromium percentage. The coatings made with this type of powders, WIP-C2 allow the subsequent processing of the surface which can be done both by milling and by turning (at the same time and/or grinding). The morphology of the powder can be seen in Figure 3.7, the chemical composition in Table 3.6 and the mechanical properties in Table 3.7 [**Error! Bookmark not defined.**].

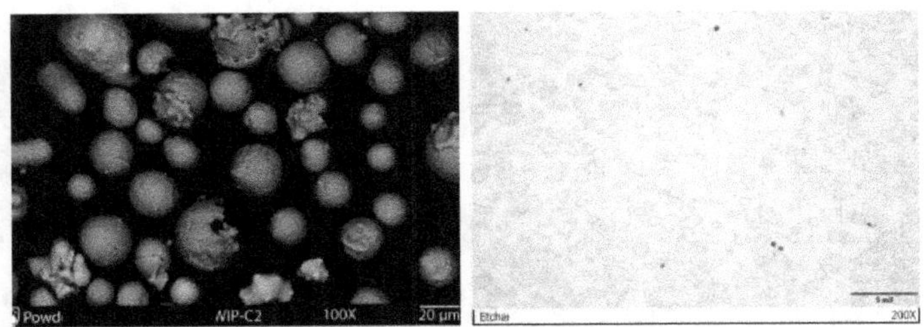

Figure 3.7 Morphology of WIP-C2 powders (a) and the sprayed surface (b)
[**Error! Bookmark not defined.**]

Table 3.6 Chemical composition of WIP-C2 powders [**Error! Bookmark not defined.,Error! Bookmark not defined.**]

Element concentration (%)	Co (%)	Cr (%)	Fe (%)	Ni (%)	W (%)	Al (%)	B (%)	Mo (%)	Si (%)	V (%)	C (%)
WIP-C2	0.01	72.01	0.08	24.35	0.01	0.07	0.04	0.01	0.39	0.01	2.94

Tabel 3.7 Mechanical properties of WIP-C2 powders [**Error! Bookmark not defined.**]

Material	Geometry	Hardness	Porosity
Carbide	Spheric	425-475 [HV]	<2% (N2)

DARC-AL6061-G1H1 powders, which are based on 6061 aluminum alloy, are thermally treated in order to eliminate harmful surface contaminants and increase the improvement of mechanical properties, at the same time, the powders have in their mixture a ceramic peening agent in order to improve the quality of the sprayed surface, by means of the cold spraying process (with nitrogen). In order to achieve a strong bonding layer between the base substrate and the sprayed layer, it is recommended to use DARC-AL6061BC-G1H1. In particular, this type of powder is designed for the dimensional restoration of various components, as well as in the repair and cosmetic processes. The morphology of the powder can be seen in Figure 3.8, the chemical composition in Table 3.8, and the mechanical properties in Table 3.9 [**Error! Bookmark not defined.**].

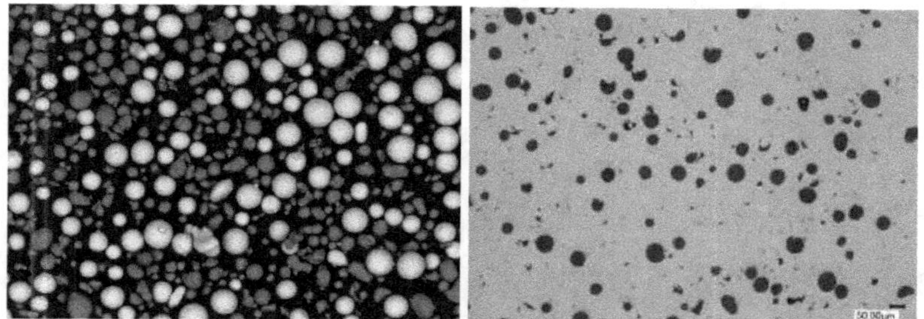

Figure 3.8 Morphology of DARC-AL6061-G1H1 powders (a) and the sprayed surface (b) [**Error! Bookmark not defined.**]

Table 3.8 Chemical composition of DARC-AL6061-G1H1 powders [**Error! Bookmark not defined.,Error! Bookmark not defined.**]

Concentration of elements (%)	Cr (%)	Cu (%)	Fe (%)	Mg (%)	Mn (%)	Si (%)	Ti (%)	Zn (%)	Al (%)	Other (%)
DARC AL6061-G1H1	0.11	0.28	0.11	0.98	0.0043	0.61	0.01	0.014	100% bal.	<0.20

Table 3.9 Mechanical properties of DARC-AL6061-G1H1 powders [**Error! Bookmark not defined.**]

Material	Geometry	Hardness	Porosity
Aluminium	Spheric	99 [HV]	<0.5% (N2)

3.3 Powders used for coatings that improve corrosion resistance

HIT-TA powders

This type of powders, used in the cold spraying process, presents special properties in order to obtain superficial layers with a very high degree of resistance to corrosion, at the same time, presenting very good characteristics when subjected to environments with high temperatures. In the cold spraying process, both nitrogen and helium can be used as carrier gas. The morphology of the powder can be seen in Figure 3.9, the chemical composition in Table 3.10 and the mechanical properties in Table 3.11 [**Error! Bookmark not defined.**].

Figure 3.9 Morphology of HIT-TA powders (a) and the sprayed surface (b)
[**Error! Bookmark not defined.**]

Table 3.10 Chemical composition of HIT-TA powders [**Error! Bookmark not defined.,Error! Bookmark not defined.**]

Concentration	Ta	O	N	H

of elements (%)	(%)	(%)	(%)	(%)
HIT-TA	100 (%) bal.	0.034	<0.001	0.001

Table 3.11 Mechanical properties of HIT-TA powders [**Error! Bookmark not defined.**]

Material	Geometry	Hardness	Porosity
Tantalum	Angular	281 [HV]	<0.5% (N2)

Metco 4016D powders

This type of powders, based on Titanium (and Titanium alloys), presents advantages through the properties that lead to obtaining a dense coating, with a low porosity and with a very good resistance to corrosion. The morphology of the powders can be seen in Figure 3.10 and the chemical composition in Table 3.12 [81].

Figure 3.10 Metco 4016D powder morphology [**Error! Bookmark not defined.**]

Table 3.12 Chemical composition of Metco 4016D powders [**Error! Bookmark not defined.**]

Concentration of elements (%)	Ti (%)	Al (%)	Fe (%)	C (%)	H (%)	O (%)	N (%)	Si (%)	Cl (%)	Mg (%)
4016D Metco	100% bal.	≤0.05	0.15	0.03	0.03	0.40	0.02	0.04	0.20	0.20

3.4 Microstructural properties of the coatings by the Cold Spray method

Cold spraying technology, through the dynamics with which it makes its way to the contributions regarding the performance of the components that are subjected to surface spraying, contributes both to the improvement of this process and to the improvement and optimization of the powders in order to obtain microstructures with superior properties that meet the needs of the industry.

In a study carried out by Jun Yan Lek and collaborators, through the cold spraying technique, coatings based on Ti-6Al-4V (Ti64) were made on a Ti-6Al-4V substrate. The microstructural evolution that starts from the powder, as a raw material, to the coating as a result of the deformation due to the high rate of plastic deformation that it presents, was investigated. Visibly, it can be observed that the microstructure is different from the microstructure of the powder and that of the coating at the interface. Near the surface (both in the substrate and in the particles), narrow regions were observed, represented by grains of a nanometric size, this remark, being concluded by the fact that an adiabatic shear instability is observed, with a localized deformation showing a high speed of deformation, together with an accumulation of heat. The drastic change in the microstructure of the particle is due to the high-velocity impact during the cold spraying process that leads to a smaller grain size (<50nm), on both sides of the particle periphery (<50nm at the periphery and 200nm at the middle of the particle), and at the upper part of the particle, away from the impact center, martensitic structures are found. As can be seen, Figure 8 (a) represents an HRTEM image where the particle-substrate interface near the particle edges can be seen. The "p" and "s" insets show the distribution corresponding to the particle area and also to the substrate area adjacent to the interface. Figure 3.10 (b) represents the HRTEM image corresponding to the binding region (the area highlighted by the square "T") in Figure 3.11 (a), with an inset showing the diffractogram corresponding to this region [82,83,84].

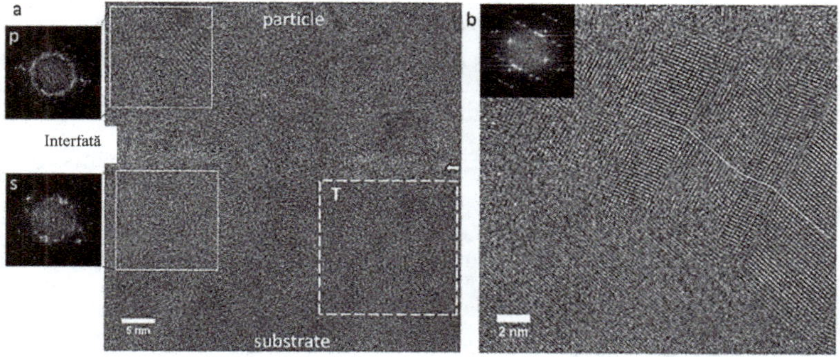

Figure 3.11 HRTEM images: (a) shows the particle-substrate interface and (b) shows the bonding area [**Error! Bookmark not defined.**]

Figure 3.12 shows schematically, following the cold spraying process, the microstructure of the particles from a coating in the vicinity of the substrate in which the fine grains are represented by small circles and the coarser grains by large circles, and the martensitic structure is represented by lines [**Error! Bookmark not defined.**].

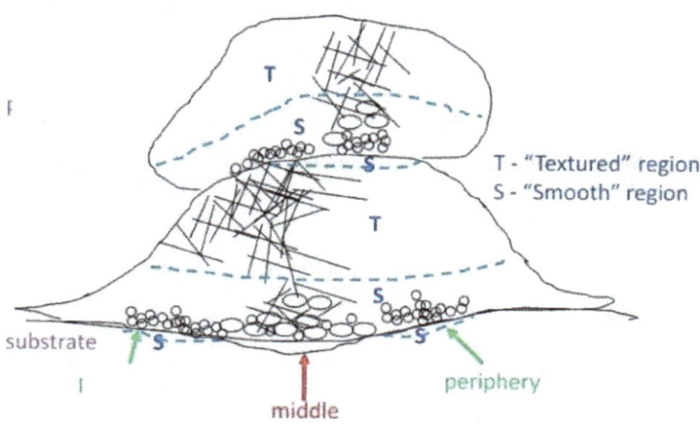

Figure 3.12 The microstructure of the particles seen with the aid of the schematic representation [**Error! Bookmark not defined.**]

At a first impression of the morphology of the deposited particles, it can be observed that, for the most part, the particles are strongly deformed and flattened. "Textured" and "smooth" regions can be observed in most of the particles, and the volume ratio between these regions is influenced by the speed and size of the particles, but also by the point of impact on previous deposits. Figure 3.13 shows an unetched cross section, where (a) represents the coating-substrate interface and (b) the inner region and "T" indicates the "textured" region and "S" indicates the "smooth" region [**Error! Bookmark not defined.**].

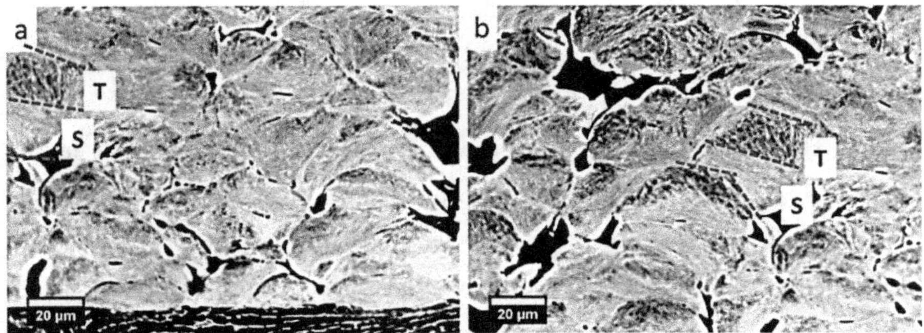

Figure 3.13 Cross section Ti64 coating [**Error! Bookmark not defined.**]

3.5 Mechanical properties of the coatings by the Cold Spray method

The cold spraying method, through the multitude of improvements it brings through surface spraying, manages to give mechanical characteristics that not only manage to satisfy needs, in terms of the restoration of various components, but also to give new parts that require surface (with a functional role or not), depending on the needs, to present a dense, very hard layer.

D. Lioma and colleagues performed a study on the hardness of deposits by low-pressure cold spraying of WC-Ni and WC-12Co-Ni powders with different percentages of Ni (4, 10, 25, 50% Ni) on a soft steel. Following the tests carried out, analyzing the SEM and XRD results, it was observed that during the formation of the coating due to the excellent refinement of the particles and the low porosity (up to 1.47%), as well as through a small percentage of Ni, unlike coatings with other values of the percentage for WC-4% Ni and WC-12Co-4% Ni, an increased hardness is obtained. From a total number of 8 combinations carried out in the study, for the WC-Ni powder, respectively WC-4% Ni, the maximum hardness value of 459.2 HV_{03} was obtained, however, this is not a satisfactory value compared to the general, average values of other coating techniques, such as CGDS and HVOF, which reach values of about 1135 HV_{03}. The microstructures with the composition WC-Ni (a, b, c, d) and WC-12Co-Ni (e, f, g, h), where the white phase represents WC, the light gray phase represents Co and the dark gray phase represents the Ni matrix, are presented in Figure 3.14, and the properties of the coatings are expressed in Table 3.13 [85].

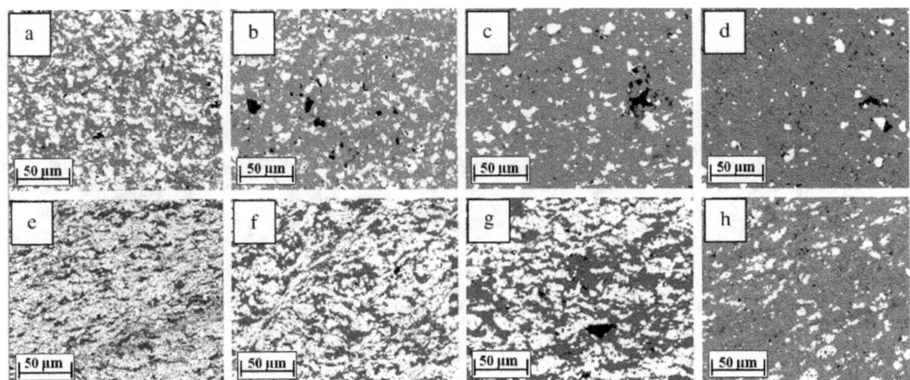

Figure 3.14 The microstructure of WC-Ni and WC-12Co-Ni powders by cold spray deposition [**Error! Bookmark not defined.**]

Table 3.13 Properties of coatings [**Error! Bookmark not defined.**]

Ni (%)	Hardness (HV$_{0.3}$)		Porosity (%)		Binder Mean Free Path (µm)		WC Retained (%)	
	WC	WC12Co	WC	WC12Co	WC	WC12Co	WC	WC12Co
4	459 ± 19	414.9 ± 23.2	1.47	3.51	5.42	1.77	39.89	74.08
10	391.5 ± 16.1	362.2 ± 32.6	2.27	3.15	12.54	3.10	24.47	63.54
25	305.6 ± 17.0	294.3 ± 31.6	2.70	3.79	21.21	4.85	17.07	47.92
50	275.8 ± 16.3	260.9 ± 25.4	2.49	5.11	48.59	10.09	6.92	27.06

NM Melendez and collaborators conducted research on the wear behavior on a low carbon steel substrate of tungsten carbide (WC) and Ni metal matrix coatings made by low pressure cold spraying, the percentage of WC of these WC-Ni coatings being 7% WC, 19% WC, 56% WC and 66% WC, at the same time performing samples sprayed with a layer of pure nickel, subsequently, the samples being subjected to standard ASTM G65 wear testing to dry abrasion. Following the analyses carried out, it was found that, on the WC-Ni samples, a significant decrease in wear is observed as the amount of WC increases, which concludes that the improvement in wear is determined by the increase in the amount of WC, and in as for the sample sprayed with pure nickel, due to the observation of the growth of the crack network, the lack of WC reinforcement proved to be harmful. The worn surface of the samples coated with 7% WC (a),

19% WC (b), 56% WC (c), 66% WC (d) and also the worn surface sprayed with pure nickel (e) are shown below (Figure 3.15 and Figure 3.16) [86].

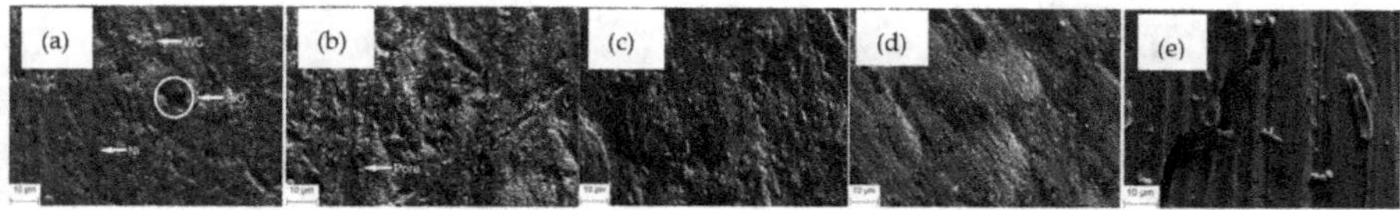

Figure 3.15 Worn surface of WC-Ni coatings (a,b,c,d) and pure nickel coatings (e) [**Error! Bookmark not defined.**]

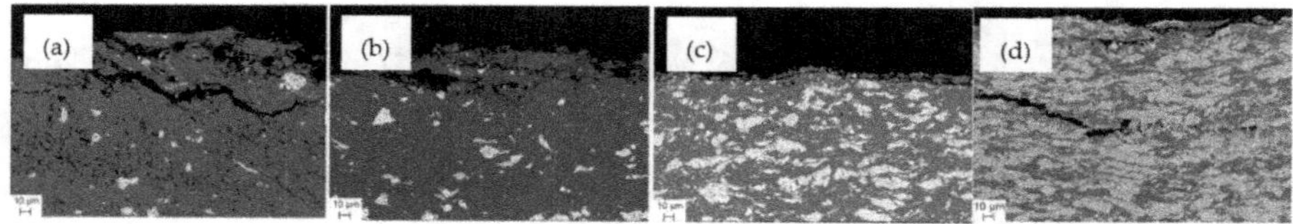

Figure 3.16 Cross section of worn WC-Ni MMC coating with 7% WC (a), 19% WC (b), 56% WC (c), 66% WC (d) [**Error! Bookmark not defined.**]

3.6 Corrosion resistance properties of the coatings by the Cold Spray method

The use of the cold spraying method and the multitude of types of powders that can be used depending on the different destinations for which they are created and optimized can create coatings that, in addition to the other properties they incorporate, manage to perfect the components that are subjected to spraying to the so-called corrosion phenomenon which, in a shorter or longer time, can produce effects that are sometimes capital for the various components so that, by superficial spraying, it is not only restored, but also given a superficial protective layer resistant to corrosion.

Fs da Silva and collaborators performed a study on the improvement of corrosion characteristics by cold sprayed WC-Co coatings on an AA 7075-T6 aluminum alloy (used in industries such as aeronautics) where due to alloying elements such as Mg, Zn and Cu, the susceptibility to corrosion is high. The tests were performed on samples coated with WC-12Co and WC-25Co. In the case of the sample coated with WC-12Co, a corrosion resistance was obtained for a time of 400 hours of immersion in a 3.5% NaCl solution, and for longer immersion times, by dissolving the cobalt phase and by losing particles of WC, led to the formation of interconnected porosities, so that the access of the electrolyte and the corrosion of the substrate were allowed, and for the other sample, namely the sample coated with WC-25Co, very good results were obtained in order to protect the AA 7075-T6 alloy against corrosion, the sample managing to exceed 700 hours of immersion in 3.5% NaCl solution, and in the case of being subjected to salt spray with neutral NaCl 5%, the sample resisted an exposure of 3000 hours. The cross section of the samples can be seen in Figure 3.17 [87].

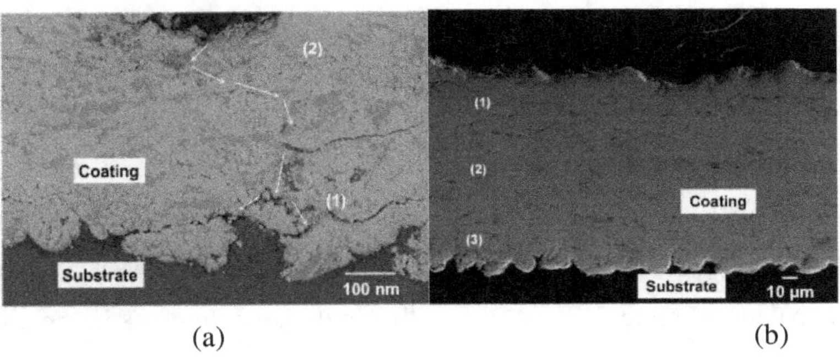

Figure 3.17 Cross-sectional image (SEM) of (a) WC-12Co coating after 600 hours and (b) WC-25Co coating after 700 hours of 3.5% NaCl immersion
[**Error! Bookmark not defined.**]

Small-amplitude linear polarization measurements taken after 18 h in a 3.5% NaCl solution were used to estimate RP values for the substrate and coated samples, and corrosion current densities were calculated using these values. The Ecorr and icorr values calculated using the Tafel plot extrapolation method from the potentiodynamic polarization curves are shown in Table 3.14 (a - the icorr values were calculated using the Stern-Geary equation, where icorr = ba/2.3 Rp and ba is the resistance to polarisation and b - values for icorr which were obtained by extrapolation from Tafel diagrams).

Table 3.14 Corrosion characteristics calculated using potentiodynamic polarisation and small amplitude linear polarisation curves after 18 h immersion in a 3.5% weight NaCl solution [**Error! Bookmark not defined.**]

Parameters	AA 7075-T6	WC-12Co	WC-25Co
$E_{i \to 0}$ (mV *vs.* Ag|AgCl|KCl$_{3\ mol/L}$	−814 ± 2	−326 ± 4	−425 ± 2
$_{rr}$ (mV *vs.* Ag|AgCl|KCl$_3$ mol/L	−831 ± 2	−336 ± 2	−425 ± 1
$^a i_{corr}$ (μA cm^{-2})	10.7	28	30
$^b i_{corr}$ (μA cm^{-2})	2.9	10	23
b_a (mV dec^{-1})	82.5	148	111

The anodic bias curves of the substrate and coating are shown in Figure 3.18 Ecorr and icorr values for AA7075-T6 were 0.8 V vs. Ag\|AgCl\|KCl$_{3mol}$/L and 2.5 Acm^2, respectively. The polarisation curve for the substrate is shown in Figure 3.18 (a). Pitting corrosion caused the current to increase when the potential reached 0.70 V vs. Ag\|AgCl\|KCl$_{3mol}$/L. The cathodic curves produced in the presence of oxygen are shown in Figure 3.18 (b). These curves revealed two current plateaus that were attributed to oxygen reduction in the range of −0.5 to −0.9 V vs. Ag\|AgCl\|KCl3mol/L and water reduction at higher negative potentials. The cathodic curves produced in the absence and presence of oxygen are compared in Figure 3.18 (c). While the predominant process in the absence of oxygen was water reduction and hydrogen evolution, the introduction of oxygen caused a rapid electrochemical reaction that was diffusion controlled [**Error! Bookmark not defined.**,88].

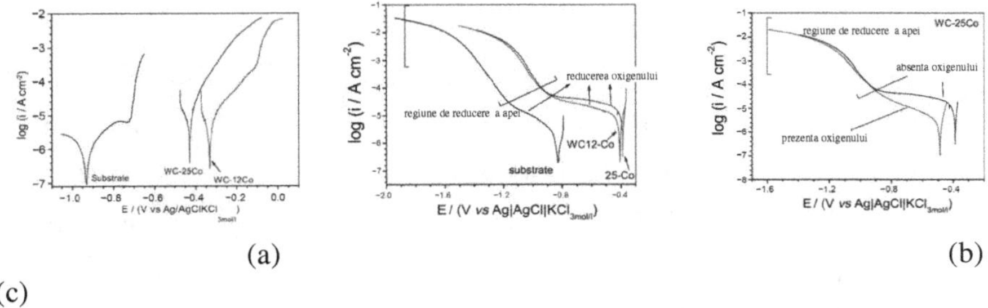

(a)

(b)

(c)

Figure 3.18 (a) Anodic and (b) cathodic potentiodynamic bias curves for AA 7075-T6 alloy and coated samples, recorded at 0.166 mV s-1 in 3.5% NaCl reagent solution after about 18 hours of immersion. (c) WC-25Co cathodic coverage curves obtained in the same solution with and without oxygen [**Error! Bookmark not defined.**]

Following a review by Ashokkumar et al. on sputtering technology and its corrosion benefits, their review concludes that the benefits of low particle temperatures in this sputtering process result in a reduction in the degree of of corrosion on the coated surface which is a high improvement compared to the other technologies (eg: HVOF and APS). From the approach of this paper on the technological process of cold thermal spraying, the idea emerges that this process has the potential to dominate the sector of thermal spraying technologies in industries such as aerospace, marine and automotive in the coming years [89,90].

In research by Zhang et al. to repair defects and prevent corrosion for AA2024-T3, thick Al-based layers were effectively coated, using the low-pressure cold sputtering approach. To create improved peening effects, spherical Al_2O_3 particles were added to the raw material. Based on the results of the cyclic polarization studies, it can be concluded that the primary form of corrosion of the tested coatings was pitting. From the analysis comparing the three depositions, the Al 2024 coating showed the weakest result, while the CP Al layer led to the strongest pitting resistance and passivation ability. According to SVET data, after 24 hours of immersion in a 3.5% NaCl aqueous solution, the

CP Al coating showed the lowest pitting activity, while the Al 2024 deposition showed the highest activity. The Al 2024 coating showed the highest corrosion rate throughout the EIS tests, in a 3.5% NaCl aqueous solution. The Al 5083 coating resulted in the lowest corrosion rate once the CP Al coating test was completed. As seen in Figure 3.22, the corrosion behaviour of cold sprayed coatings was investigated. According to Figure 3.19 (a), the pitting location on the CP Al coating surface is substantial, deep, and uneven in shape. A spherical Al_2O_3 particle is found at the edge of the crack, and no preferential dissolution of the Al matrix is shown at the $Al-Al_2O_3$ interface, which shows that no galvanic cell has formed between the Al matrix and the Al_2O_3 particle. In addition, a layer of corrosion products has developed on the matrix, which can partially isolate the electrolyte layer. On the surface of the Al 2024 coating, there are several cracks of relatively small depths that are randomly positioned, as seen in Figure 3.22 (b). When comparing the cracks on the Al 5083 coating with those on the CP Al coating, Figure 3.22 (c), it demonstrates that the Al 5083 coating cracks are shallower and smaller in size [91].

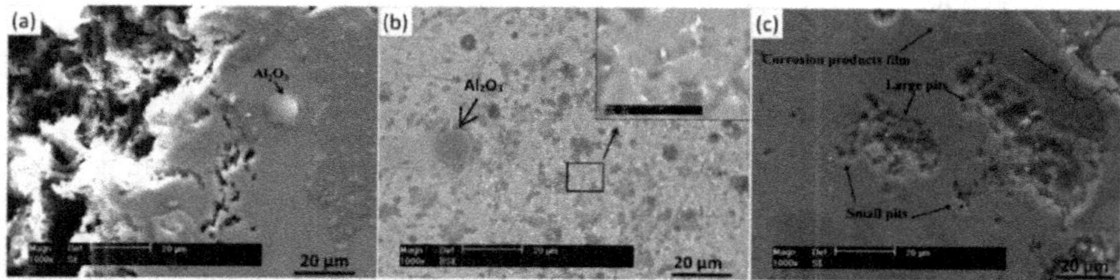

Figure 3.19 Surface morphology of CS coatings following CP corrosion test: (a) represents Al coating, (b) represents Al 2024 coating and (c) represents Al 5083 coating [**Error! Bookmark not defined.**]

Figure 3.20 shows the cyclic polarization curves. Possible scan routes are shown by the solid arrows adjacent to the forward and reverse anodic branches. At a certain critical potential, known as the Epit pitting potential, where a sudden increase in anodic current occurs as a result of the breakdown of the passive film and the initiation of pitting corrosion, the corrosion currents of Al and Al 5083 coatings decreased rapidly with increasing potential and respectively, they reached the minimum at Ecorr. However, when the potential increased at Ecorr, the corrosion current of the Al 2024 coating decreased at a very slow rate. The current increased immediately after that. Epit can be a useful option for showing corrosion resistance, as higher pitting potential often equates to better pitting corrosion resistance [**Error! Bookmark not defined.**,92,93].

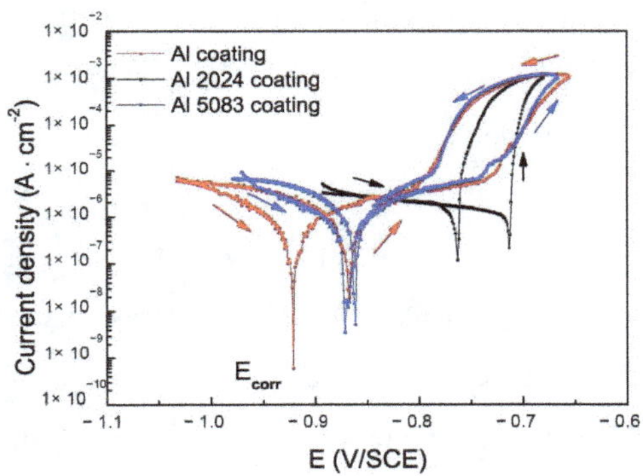

Figure 3.20 After 6 hours of immersion in a 3.5% NaCl solution, the cyclic polarization of the coated samples was measured at 1 mV/s [**Error! Bookmark not defined.**]

3.7 Results obtained so far by using the Cold Spray method

Cold spraying technology, through the fact that it succeeds in improving the properties of the materials leading to an improvement in the characteristics and functionality of the various components that are subject to this process and, more than that, through the possibility of depositing on a wide range of materials such as metals, ceramics, polymers, etc. This technique manages to gain value and be used in numerous industries such as: the energy sector in order to extend the life of boilers damaged by corrosion and oxidation from the high temperatures of thermal power plants; the aerospace industry where cold spraying manages to restore and repair numerous defective parts that are rejected in the technological manufacturing process because they present very high quality standards, and this technology manages to return the parts to the desired characteristics; the biomedical sector where cold spraying manages to lend itself through the possibility of depositing on different types of orthopedic and dental implants through coatings, (for example, with hydroxyapatite and titanium (ceramic coatings)) that improve osseointegration, etc. [79,80,81].

In a study by Yongming et al. on HEA composite coatings, in which FeCoNiCrMn HEA composite layers reinforced with Al_2O_3 (11 and 20% Al_2O_3) made by cold spraying, showed that cold spraying is a potential approach for creating HEA composites coatings. The interface between the substrate and the coatings, as applied, can be seen in Figure 3.21 (a - c (SEM images)). It can be seen that all coatings demonstrated strong adhesion to the aluminum alloy substrates, as evidenced by the smooth interface free of voids and fractures. Along the contact between the substrate and the coating, a mechanical, interconnecting phenomenon was visible. The substrate material firmly gripped certain particles that managed to penetrate it. When the hard FeCoNiCrMn particles strike the soft substrate material, there is strong plastic deformation that leads to the development of mechanical interlocking. A prominent feature of cold sprayed coatings is mechanical interlocking, which can give the coatings strong adhesion. Cross-sectional SEM images of the deposited coatings are shown in Figure 3.21 (d - f). The FeCoNiCrMn coatings clearly show that they have a dense microstructure with fewer pores, as indicated by the yellow arrows and the particle boundaries indicated by the red arrows. As seen in Figure 3.24 (e,f), the composite coatings also exhibit a coarse microstructure. The white arrows indicate where, in the FeCoNiCrMn matrix, the Al_2O3 particles were uniformly distributed. Red arrows indicate voids between particles, increasingly visible for higher percentage of Al_2O_3 [97,98].

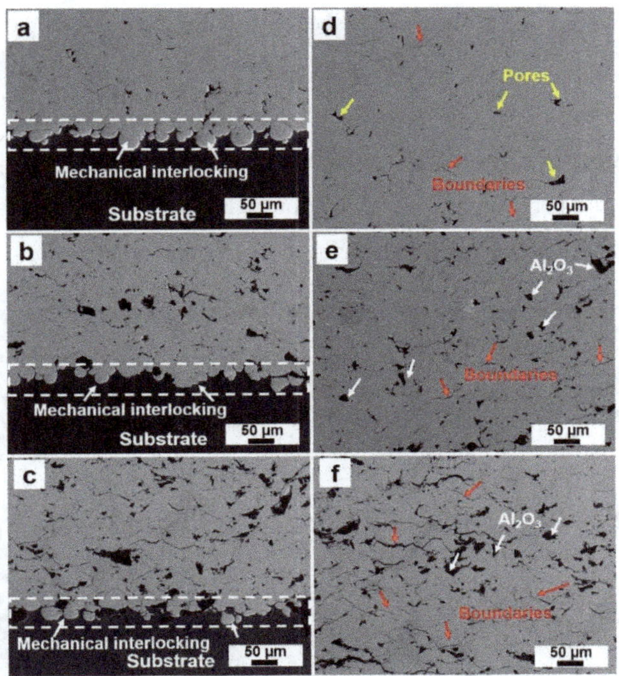

Figure 3.21 SEM images of coating-substrate interfaces and coatings in cross section. The pure FeCoNiCrMn coating is shown in (a,d), while (b,e) shows an 11% Al$_2$O$_3$ coating and (c,f) a 20% Al$_2$O$_3$ coating [**Error! Bookmark not defined.**]

Aaron Nardi and collaborators from the U.S. Army CCDC Research Laboratory (also known as Devcom) studied the effects of cold spray coatings, to restore and improve the quality of various military technical components, using WIP powders. Several types of powders were used in the study, namely, WIP-BC1 powders as the first base layer (bonding layer), followed by the second WIP-C1 powder, and the component that was studied for spraying, being made of alloy steel AISI 4340. Research findings led to similar and even better wear performance than chromium (Cr) plating, increasing surface quality for high impact conditions, and shear tests led to an increase in hardness from 17 HRC of the base material up to 40-44 HRC, and the surface porosity resulting from sputtering was 1.5-3%, in the case of nitrogen (N2) sputtering and below

1% using helium (He). The cross-section of the layer made with WIP-C1 powders, as well as of the sprayed base substrate (binder layer), with WIP-BC1 powders, can be seen in Figure 3.22, and the part that was subjected to spraying where the superficial layer is observed, is exposed in Figure 3.23 [99]. Figure 3.24 shows another example of deposition where successive depositions were made to achieve a relatively large thickness of the deposited layer.

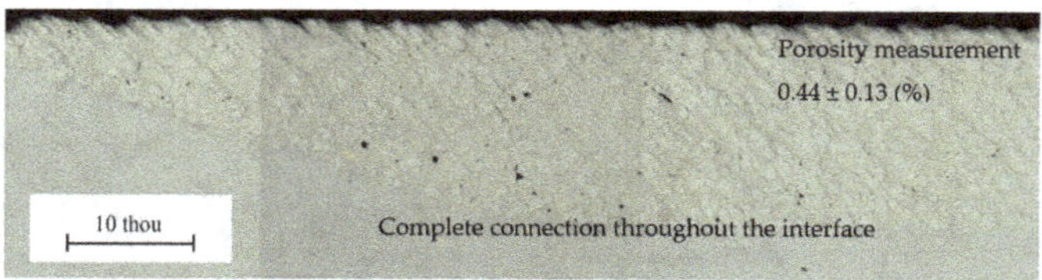

Figure 3.22 Sprayed surface using WIP-BC1 powders and WIP-C1 powders (1 thou=0.0254 mm) [**Error! Bookmark not defined.**]

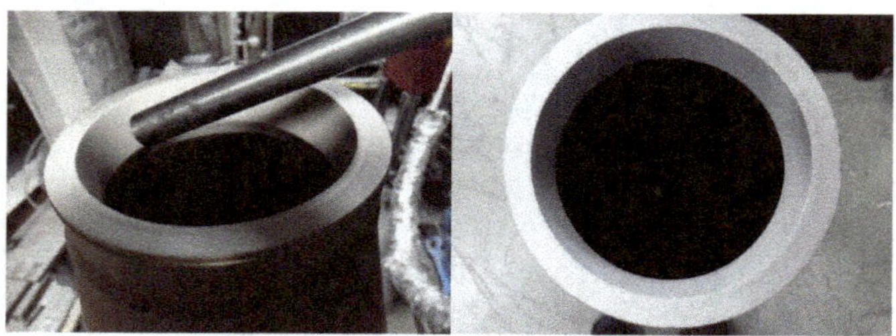

Figure 3.23 Turret mount (Bradley) cold sprayed [**Error! Bookmark not defined.**]

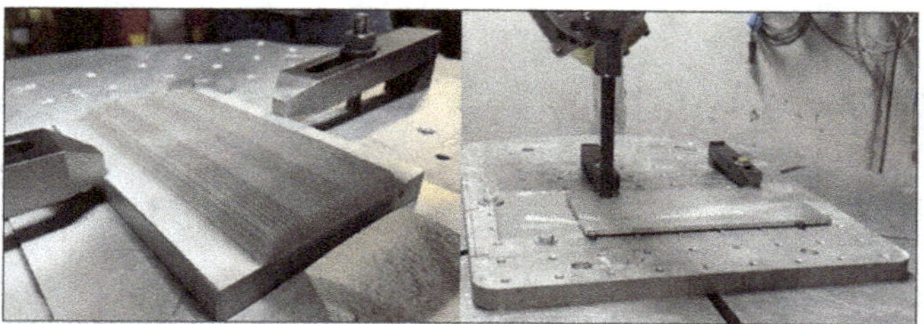

Figure 3.24 Surface resulting from deposition by the Cold Spray method [**Error! Bookmark not defined.**]

In an article published by the US Federal News Program (Phillips), Army materials scientist Dr. Victor Champagne, who in 2001 established the Cold

Spray Center, at the US Army Research Laboratory, claims that the technology of spraying (cold spraying) is a revolutionary one, due to the fact that the improvement of powders led to the spraying of almost impossible materials. This technology proves to be quite effective in terms of the technological processes of repairing various components, so it causes a continuous development of both the spraying technology and the powder used in this process [100].

The effectiveness of this technology, in terms of technological procedures for the repair of various components, leads to the continuous improvement of both the spraying technology and the powder used in this process.

In a research by Hanqing et al., SnZn and SnBi powders with lower melting points than Sn were cold sprayed onto polymer substrates (ABS, Nylon, PC, PE and PP). Due to a higher degree of melting of the particles, the results demonstrate that lowering the melting temperatures of raw material powders can increase the efficiency of their deposition on polymer substrates. Polyethylene and high-strength, extremely high molecular weight polycarbonate were two of the five polymers that were the most difficult to metallize. Metal coatings on polymer substrates are shown in cross sections, in Figure 3.25. Cold sprayed SnZn coatings at 300°C and 80 psi are shown for ABS, Nylon and PP. For UHMW and PC, SnBi coating at 300°C and 42 psi shows that SnZn deposition was insufficient or irregular on these latter two polymers. The two-phase microstructure that is present in the raw material powder is still present in all coatings which are all fairly consistent in thickness and density. As the spherical particle can be observed penetrating the substrate for SnZn coatings on ABS, nylon and PP, mechanical interlocking can be observed. The combined impact of temperature and glass transition strength can be used to explain why the degree of interfacial waviness is on the order of PP-ABS-Nylon. Ease of penetration and interlocking increases with decreasing resistance. Since mechanical interlocking and particle penetration are caused by the low content of Sn and SnZn, for the two high hardness polymers UHMW and PC, it is understandable.

The top surfaces of the substrate are comparatively smooth when UHMW and PC are metallized with SnBi, indicating that there is less mechanical interlocking (details of the SnBi-PC interface are shown by the high-magnification inset). The coating on the UHMW actually shows delamination. Extensive particle melting and low gas pressure at 300 °C prevent mechanical anchoring for SnBi [101,102,103,104].

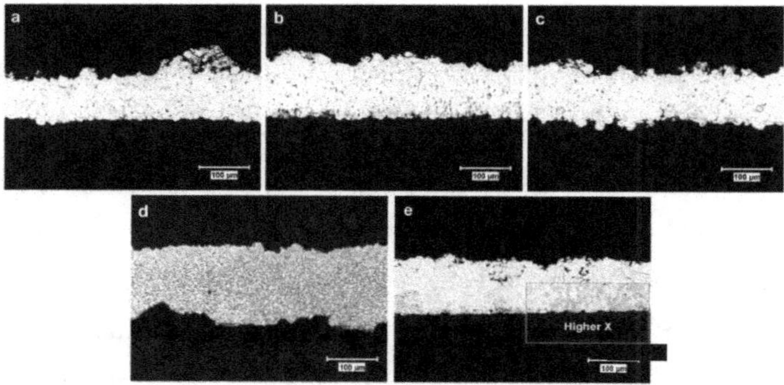

Figure 3.25 Cross-sections of cold-sprayed SnZn ((a–c) 80 psi) and SnBi ((d,e) 42 psi) coatings on all polymers at 300 °C are shown in optical micrographs: (a) ABS; (b) nylon; (c) PP; (d) UHMW and (e) PC [**Error! Bookmark not defined.**]

3.8 Partial conclusions

1. AISI 52100 steel is a chrome alloy steel with a high carbon content and is excellent for the construction of highly stressed components that require superior wear and fatigue resistance properties.

2. The powders used as raw material can be found in a fairly large range, they being specially designed for cold depositions. Depending on their chemical composition, the microstructural, mechanical and corrosion properties can be improved.

3. The powders with the commercial name of WIP-C1 and WIP C2 are Ni Cr/C based powders that allow the creation of deposits with significant thicknesses, dense and closely related to each other, managing to achieve hardness values of up to at 43 HRC.

4. The research carried out in this field confirms the superior properties of the layers deposited by Cold Spray, which are closely related to the base material, without the risk of exfoliation.

CHAPTER 4 Methods and facilities for analysis of coatings made by the Cold Spray spraying method

4.1 Cold Spray deposition plant

The cold spray system (Cold Spray) [44] with the elements that compose it, is represented schematically in the following figure (Figure 4.1.), in order to be able to understand more easily how the subsystems and elements that compose the installation are arranged:

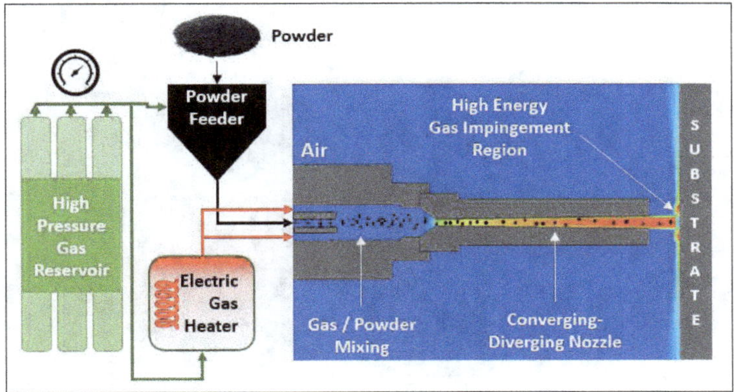

Figure 4.1 Schematic view of the Cold Spray deposition plant [44]

Cold Spray installation systems:

(a) Nitrogen generation system

The nitrogen generation system is composed of an air compressor, oxygen filter membranes and a booster compression. Nitrogen is stored in a bank of gas cylinders.

Figure 4.2 Nitrogen generation system [44,105]

(b) VRC Gen III cold spray system

This system allows for cold spray depositions, maintaining the pressure and temperature of the inert gas stream, and, at the same time, a constant rate in terms of powder feed.

Figure 4.3 VRC system [44,**Error! Bookmark not defined.**]

(c) VRC powder feeder

The feeder assembly is characterised by its disposition in an inclined position and by a rotary dosing device, which allows constant maintenance of the advance speed.

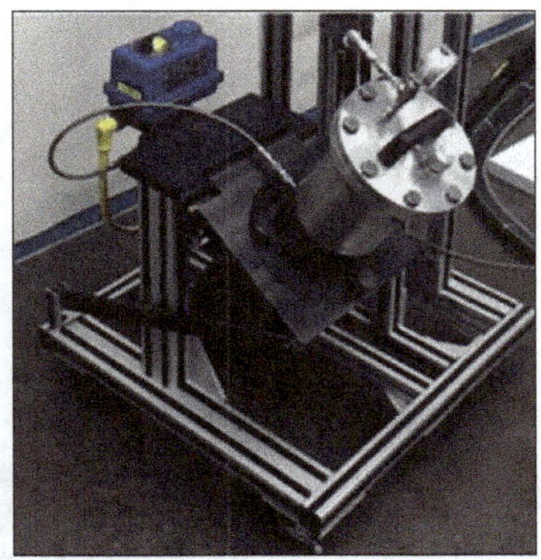

Figure 4.4 VRC powder feeder [44,**Error! Bookmark not defined.**]

(d) VRC block applicator

Before the heated gas and powders reach the nozzle, they pass through a system of pipes that join in the applicator, where the mixing is done.

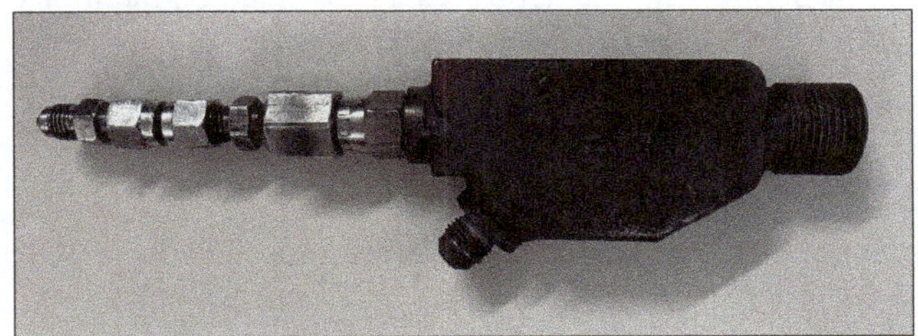

Figure 4.5 VRC block applicator [44,**Error! Bookmark not defined.**]

(e) VRC cold spray nozzle

This is a convergent-divergent type nozzle that allows the carrier gas to be accelerated to supersonic speeds.

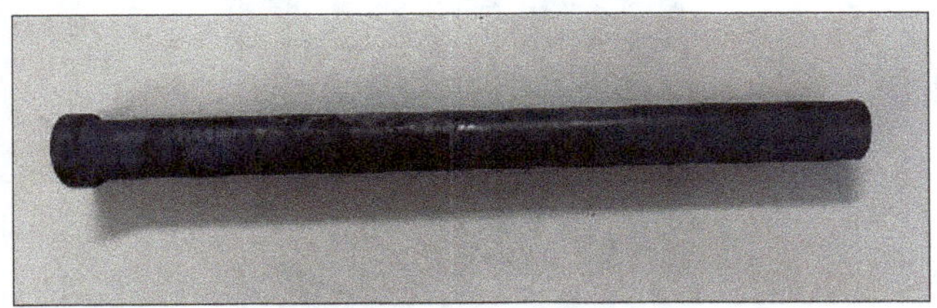

Figure 4.6 VRC spray nozzle [44,**Error! Bookmark not defined.**]

(f) Fanuc robot and rotary table

The applicator and nozzle are mounted on a robot arm. The robot is programmed so that the areas intended for deposition are covered with cold spray.

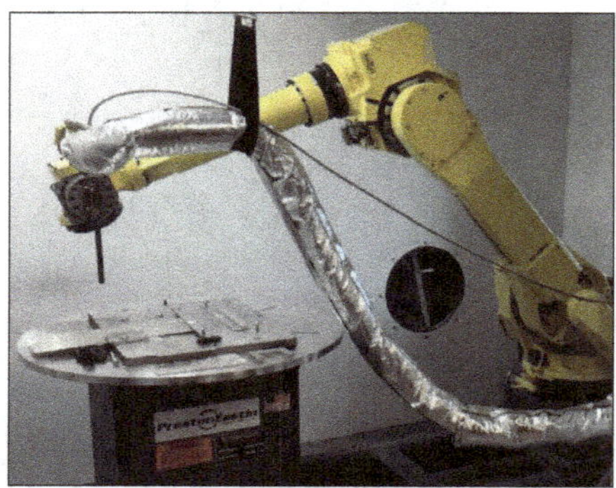

Figure 4.7 Robotic arm (the deposited part is fixed on the rotary table) [44,**Error! Bookmark not defined.**]

(g) Dust collector

Metal dust particles are extracted from the spray chamber, with a wet dust collector operating at 3,000 cfm.

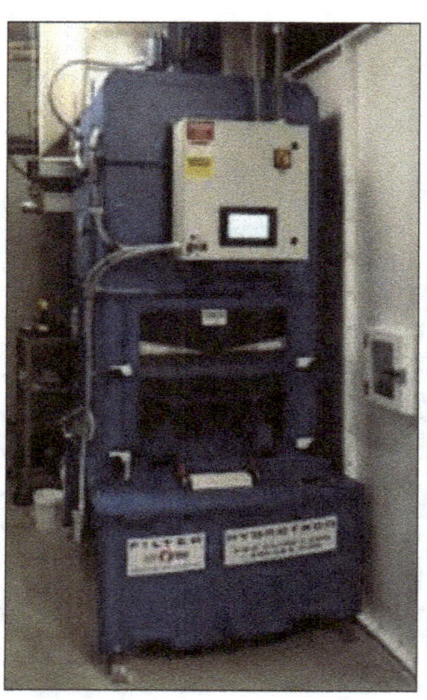

Figure 4.8 Dust collector [44,**Error! Bookmark not defined.**]

(h) Acoustic enclosure

The carrier gas coming out of the nozzle is extremely powerful, so the spray booth is inside a secondary sound enclosure.

Figure 4.9 Work room [44,**Error! Bookmark not defined.**]

4.2 Equipment used in structural analyses

4.2.1 The Quanta 200 3D Dual Beam scanning electron microscope

Due to the fact that it can perform extremely complex analyses, the scanning electron microscopy is increasingly used in fields such as: engineering, medicine, physics, chemistry, etc., thus managing to be a vital tool in high-performance research.

Electron microscopy allows research to be carried out by performing at least the following analyses, as follows [106]:

- studies on the morphological properties of metal surfaces (allows both the observation of the structure and the orientation of electrodepositions and oxides);

- allows the analysis of the metal surface regarding the stability of the properties of the oxide structure;

- studies on chemical metallurgy processes;

- allows the study of massive samples, both iron and other ferromagnetic alloys, regarding their magnetic domains;

- allows observing the influence of substances on metal surfaces such as: water vapor, salt fog, etc.).

- by means of the correlation between the spectrophotometer and the electronic scanning microscope equipped with a computer, it allows the stabilization of the chemical composition of the sample under analysis, as well as the stability of the composition from a qualitative and quantitative point of view;

- allows the observation of the phenomena that occur at the surface level, both for metals and for other metals, as follows: cleavage fracture, fatigue fracture, studying the internal structure of the sample, observing simple non-brittle deformations, observing fracture surfaces (and the rough ones), etc.

- allows the study of the oxidation of metals and alloys, at high temperatures with regard to structural and morphological examination;

- allows the study of phenomena such as chemical transport inside the structures of metals and metal alloys;

- allows studying the structure and also the distribution of the cavities that form at the surface level of the metal-metal oxide.

The Faculty of Mechanics within the "Gheorghe Asachi" Technical University of Iaşi has a Quanta 200 3D Dual Beam Scanning Electron Microscope, produced by the company Fei (Field Electron and Ion Company) [107], in the Department of Materials and Surface Engineering, with a significant impact on the manufacturing market of electronic microscope systems, whose origins are in the United States, having work points also in countries such as France, Germany, Holland, etc.

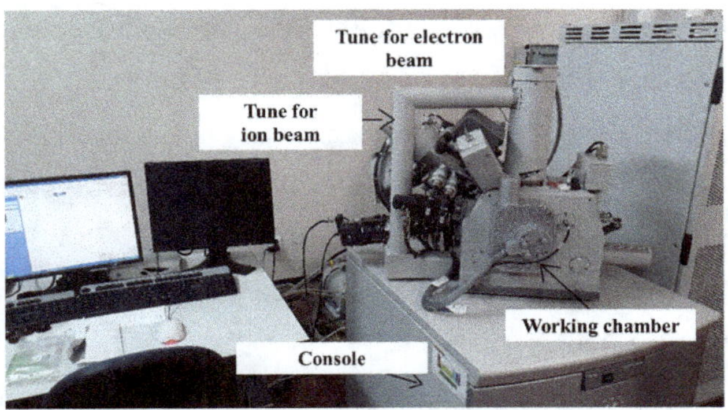

Figure 4.10 Scanning electron microscope Quanta 200 3D Dual Beam, provided by the Faculty of Mechanics, "Gheorghe Asachi" Technical University of Iaşi [108]

The systems (workstations) that make up the Quanta 200 3D Dual Beam electronic microscope, are the following [**Error! Bookmark not defined.**,109]:

1. Scanning Electron Microscope (SEM) that allows a magnification power of up to 3,000,000X, being compatible for analysis of a wide variety of samples.

2. Focused Ion Beam (FIB) is a system that allows digging into samples (on the order of microns), by means of an ion beam that allows cross-sectional structural analysis of the superficial layers of the examined sample.

3. Energy Dispersive X-Ray Spectroscopy (EDAX) is an acquisition system that allows qualitative and quantitative chemical analysis of the elements in the chemical composition of the sample material, as well as crystallographic analyses.

Depending on the sample under examination, the software ensures the operation of workstations (FIB/SEM) in 3 modes, as follows [**Error! Bookmark not defined.**,**Error! Bookmark not defined.**]:

• Environmental Scanning Electron Microscope (ESEM), through which it is possible to examine samples that are not compatible with vacuum (for example, such as samples containing water, hydrogels, etc.).

To study the sample in its natural state, an analysis-specific gas and pressure environment is created by means of the ESEM mode.

FIM/SEM has the advantage that it can carry out studies at a much higher level, unlike other systems that do not allow them to be carried out, such as for example the following:

- captures very high resolution images of the FIB section by electron beamS, without eroding the studied features.

- During FIB milling, it allows both real-time viewing and video recording of electron beam sections.

- during FIB grinding, it allows to neutralize the charge of the electron beam.

- allows the elemental microanalysis of cross-sections to be performed at a high resolution.

- allows scanning with electron beams, without eroding or implanting the gallium in the ion beam.

- presents the possibility of conductive coating in situ, for the preparation of TEM samples.

• High Vacuum-HiVac is the normal (conventional) mode of operation of scanning electron microscopes, working both with conductive samples and with samples conventionally prepared for examination (by metallization).

• Low Vacuum-LowVac, through which both unprepared (non-metalized) samples and samples from non-conductive materials (polymers, fiber, etc., with an electrical resistance >1010 [ohm]) can be analyzed. At the same time, this mode allows performing analyses in which there is the possibility of altering the sample by the deposited superficial layer.

The components of the Quanta 200 3D Dual Beam electron microscope are as follows [**Error! Bookmark not defined.,Error! Bookmark not defined.**,110]:

- the electron or ion gun, which has as component elements an anode, a cathode and a Wehnelt cylinder.

The emission of the beam of electrons or ions is carried out in a small volume presenting, by means of the selectable energy, the possibility of angular scattering of the beam.

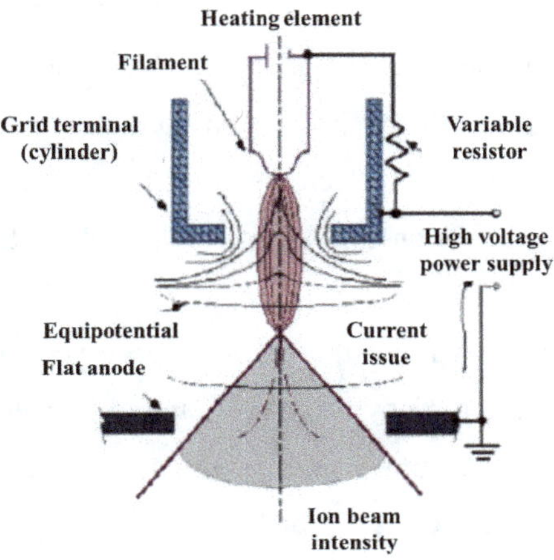

Figure 4.11 Presentation of the principle of operation of the electron gun [82]

- lens system where, through the electromagnetic lens system, the beam enters and then exits the device and hits the sample analysis surface. Unlike the fixed focal length that characterizes the other systems, through the excitation current that circulates through the coils, and the magnetic lenses allow a variable and controllable focal length. Similar to the rest of the optics, magnetic lenses offer aberrations such as: chromaticity, sphericity and diffraction.

- scanning unit that, through the deflection system, passes the scanning signal and moves the beam on the sample analysis surface, through a raster pattern. By means of this model, which changes the electrical voltage, serial information about the surface of the sample is transmitted. Modulation, through a detection system, allows the signal to be processed to obtain the image on the screen.

- detection unit, where the reaction of the atoms of the surface of the sample with the particles hitting it is different (production of signals by the electron beam, respectively: X-rays, electrons and photons; production of ions and electrons by the ion beam).

- detection system, which allows the particles or signals picked up to be converted into a signal (amplified), in order to be sent to the control computer and displayed on the monitor.

- the vacuum system, which allows the optimal operation of the electron gun filament and, at the same time, the undisturbed assurance of the electron beam that is carried out through the optical system to the sample, so that it achieves a limitation of the scattering of electrons on the atoms of the elements in the atmosphere of the column and of the chromatic aberration, which limits the resolution.

Analysis of samples without the need for prior preparation, is possible by scanning electron microscopy if the sample material allows electron bombardment and the vacuum inside the microscope [**Error! Bookmark not defined.,Error! Bookmark not defined.**].

Without prior metallization of the sample, the Quanta 200 3D Dual Beam microscope allows analysis, having set the working mode (Low Vacuum and ESEM) that presents a reduced vacuum.

For the analysis, however, it is necessary that the dimensions of the sample are not large, because they can cause difficulties when introducing it into the premises, and, at the same time, the magnetic properties of the sample can influence the analysis.

Depending on the nature of the research carried out, the samples will be properly prepared for their performance. For example, surface polishing is necessary to be able to observe inclusions or phases (with atomic number contrast), observe magnetic domains (with magnetic contrast), observe the orientation of crystal grains (grooving contrast), etc. Due to the fact that the strongest contrast is topographic, it must be removed for the best results of this case, so the need for polishing for sample preparation is extremely necessary. Just as it is done in the case of optical microscopy and in the case of electron microscopy, the metallographic study involves the use of polished and chemically attacked surfaces [**Error! Bookmark not defined.**].

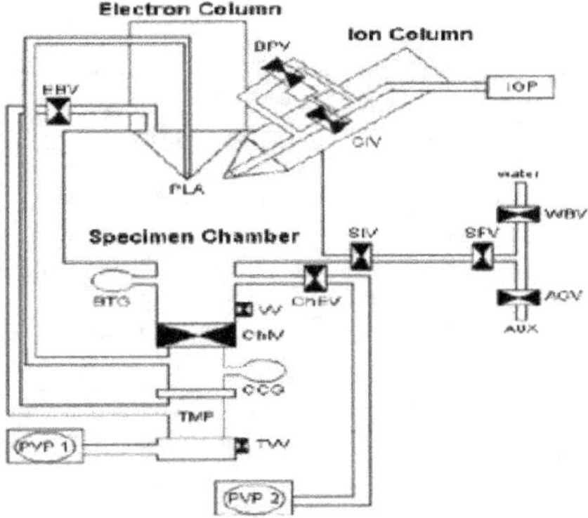

Figure 4.12 Schematic view of the vacuum system within the Quanta 200 3D Dual Beam Scanning Electron Microscope [**Error! Bookmark not defined.**]

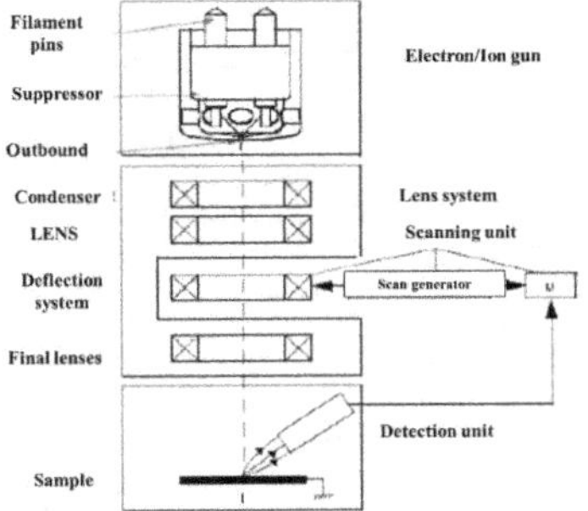

Figure 4.13 Schematic illustration of how the scanning electron microscope works [**Error! Bookmark not defined.**]

The characteristics related to elemental chemical analysis (EDX) [**Error! Bookmark not defined.,Error! Bookmark not defined.,**112] are presented as follows:

Energy-Dispersive X-ray spectroscopy (EDS/EDX) is a system that, by means of X-rays, allows the performance of elemental microanalyses. The system has the possibility of detecting changes in the surface of the sample, being able to perform both elemental chemical analyses and chemical characterization of the samples (the analysis allows the study for a depth resolution of 0.2-2µm).

EDS/EDX leads to X-ray fluorescence spectroscopic analyses where the elements that are the basis of the sample analysis represent the interaction between the sample and the electromagnetic radiation, thus being able to analyse the X-rays, as a result of the response resulting from the impact with electrical charged particles. X-rays allow the characterization of elements through the fundamental principle that attests the fact that each chemical element has a unique atomic structure.

4.2.2 Optical microscopy

In most of the research carried out in materials engineering, optical microscopy is used in a significant percentage, being a vital element of this field. Using optical microscopy, we can perform different researches, as follows **[Error! Bookmark not defined.,Error! Bookmark not defined.,Error! Bookmark not defined.]**:

- Studying samples from metals or non-metals regarding the observation of inclusions such as: puffs, nitrides, oxides, etc.;

- Establishing the shape, size and distribution of the structural constituents of the alloys;

- After carrying out the thermal and thermochemical treatments of the various components, through optical microscopy, it is possible to study their structure with regard to the defects that appeared as a result of the treatment, the thickness to which the structure changed as a result of the treatment, the new phases formed, etc.;

- Observation of crystalline grains, in the case of research carried out regarding their plastic deformation (for example, components subjected to the rolling technological process, such as sheet metal, pipe, etc.);

- Morphological analysis of the surfaces of the components;

- The possibility of carrying out research on the transformations in the solid phase resulting from the thermal influence.

The metallographic microscope is used for researching metallographic structures. The principle underlying the operation of the metallographic microscope is the reflection of light on the surface of the sample, which requires proper preparation: ensuring the flatness of the analysed surface; grinding followed by polishing and chemical attack (reagents appropriate to the sample material). The main research carried out using this type of microscope are: structural granulation, inclusions, nature, size, shape, distribution of structural constituents [113].

The main components that make up the metallographic microscope are [114]:

- the optical system, composed of: eyepiece, objectives (+filters), prisms, mirrors.

- the lighting system, composed of the light source (+ various prisms, diaphragms, lenses, filters).

- mechanical adjustment system, which ensures the movements of the table by means of a screw-nut mechanism and, at the same time, through the same type of mechanism, the movement of the objectives is also ensured.

- photo/video system, ensures both real-time capture of images and the possibility of saving them.

To carry out the analyses, type IM and DM microscopes are used, found in the equipment of the Materials and Advanced Materials Study laboratory within the Faculty of Mechanics, "Gheorghe Asachi" Technical University of Iași.

Figure 4.14 Metallographic microscope (IM 7100). The main component elements [**Error! Bookmark not defined.**]

4.2.3 X-ray diffractometer

In X-ray diffraction research, the crystal structure of the sample material can be determined, determining the symmetry and size of the lattice, and, at the same time, regarding the unit cell, the place occupied by the atoms in it can be determined. X-ray diffraction has the advantage of being able to perform studies, without damaging the component being analysed as a sample, thus being a non-destructive method of analysis. The imperfections of the crystal lattice, as well as the size and shape of the scattering domain, are analysed by means of X-ray diffraction and after performing any analysis, it is possible to obtain the results through a diffractogram.

The main studies aim at the analysis of crystal lattice imperfections, phase identification, crystallographic texture, internal stresses, measurements at different temperatures, etc.

By means of this method of analysis, studies can also be carried out on powders that are not characterised by a crystalline structure, but by a regular molecular structure, at the same time, being able to identify chemical compounds and determine the size of ultramicroscopic particles. X-ray spectroscopy leads to the determination of chemical elements and their isotropes [**Error! Bookmark not defined.**,115].

Bragg's law is the basis of the study of X-ray crystals, defining the distance between two crystallographic planes.

X-ray diffraction can be divided into two categories [**Error! Bookmark not defined.**,116]:

- The method of the 4 circles and the Laue method (study of monocrystalline materials);

- Debye-Scherrer method and diffractometer method in Bragg-Brentano geometry (study of polycrystalline materials).

The sample subjected to analysis allows the existence of a flat surface, the X-ray being monochromatic, it is possible to analyse both compact material and powder samples, and the digitization of the analysis equipment raises the study to a high level by controlling the parameters, through the software and the computer the diffractometer [Error! Bookmark not defined.].

Available for the analysis, the X-ray, X'Pert PRO MRD diffractometer, is available in the equipment of the Materials Study and Advanced Materials laboratory of the Faculty of Mechanics, "Gheorghe Asachi" Technical University in Iași. It is manufactured by the Dutch manufacturer, Malvern Panalytical [117].

With the help of this device, it is possible to carry out the following analyses, as follows:

- powder analyses;

- crystallographic analyses (on different samples with different types of surfaces, such as flat or irregular ones);

- analyses involving the quantitative/qualitative assessment of different coatings;

- analyses involving the quantitative assessment of microtensions;

- analyses regarding the determination of matrix deformation and crystallite size;

- analyses on the texture of the material;

- analyses on the nanopowders regarding the study of the dimensional distribution, as well as the surface area, etc.

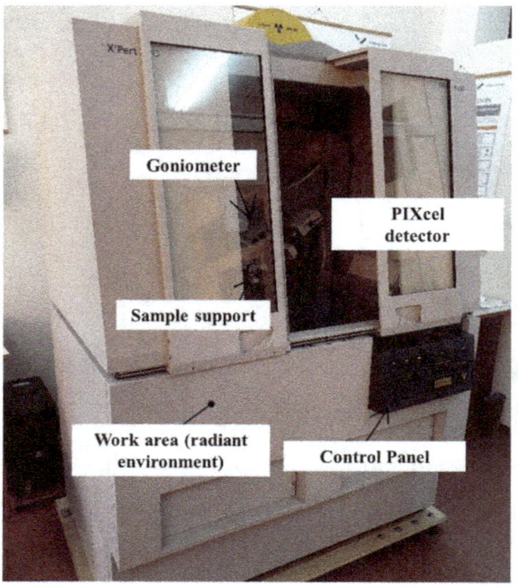

Figure 4.15 Introducing the X-ray diffractometer, Panalytical X'Pert Pro [**Error! Bookmark not defined.**]

The components of the X'Pert PRO Diffractometer are as follows [**Error! Bookmark not defined.,**Error! Bookmark not defined.]:

- work area (console);

- the goniometer which represents the main element of the assembly;

- X-ray ceramic tube (Cu anode) equipping an arm of the diffractometer;

- cooling system, designed to maintain the X-ray tube at the optimal operating temperature;

- optical modules for X-rays (incident and diffracted sample);

- sample support adaptable according to the sample and the measurements performed;

- detector, which analyses the X-ray beam at the parameters specific to the analysis. Through the Medipix 2 technology, the detector is the X-ray detection assembly, with an efficiency of 94%, being built to work with copper radiation, but also with other types of radiation, but not at the same efficiency.

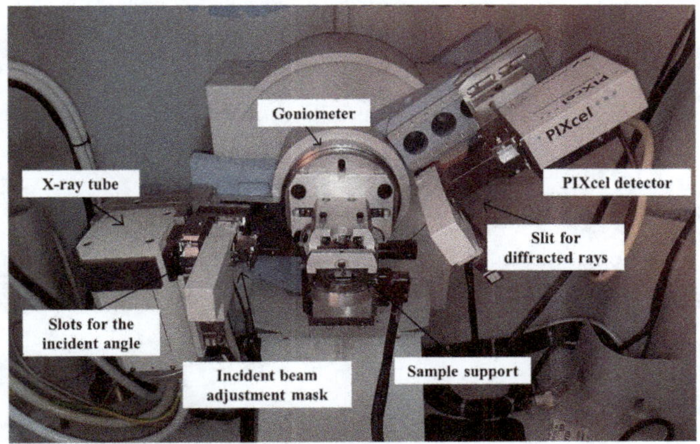

Figure 4.16 The main elements of the goniometer [**Error! Bookmark not defined.**]

The main elements that make up the goniometer assembly are:

- X-ray tube;

- PIXcel detector;

- slot for the incident angle;

- slot for diffracted rays;

- incident beam adjustment mask;

X-ray tube:

As it can be seen in Figure 2.8 (b), it is composed of an anode and a cathode (in a shell), integrated in a focusing cylinder provided with holes for the release of X-rays.

The X-ray tube includes the following technical features:

- maximum power of 2.2Kw;

- maximum voltage of 60 KV;

- maximum intensity of 55 mA;

- yield of 0.2%

- operation at maximum performance can be obtained through the following configuration: 40KV, 45mA.

X-ray tubes are found with the anode made of a variety of materials such as: Co, Fe, Mo or Cr, the choice of material from which the anode is made, being determined by the nature of the samples under analysis.

Depending on the analysis being carried out, as well as the size of the sample used and the desired accuracy of the measurements, the X-ray tube allows emitting in 2 modes, respectively line and point (e.g. "line-focus" position).

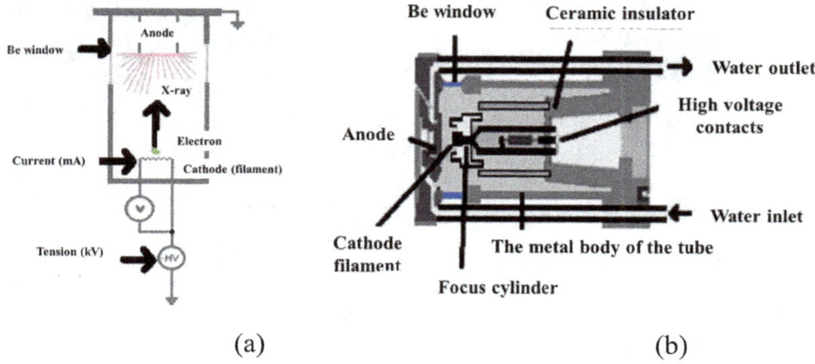

Figure 4.17 X-ray tube. Principle of operation (a); section through the x-ray tube (b) [**Error! Bookmark not defined.**]

In terms of performing analyses with a clarity of the obtained information, the X-ray beam is a wide one to capture the information and in order not to reflect, it is composed of a series of slits that narrow the beam.

The PIXcel detector

The X'Pert PRO MRD diffractometer has in its assembly, an essential component, called the Detector. This component is built and parameterized to work with copper radiation, its efficiency being 94% in this case and in the case of using other types of radiation, it decreases. The technology used by the detector is Medipix 2 [**Error! Bookmark not defined.,Error! Bookmark not defined.,Error! Bookmark not defined.**].

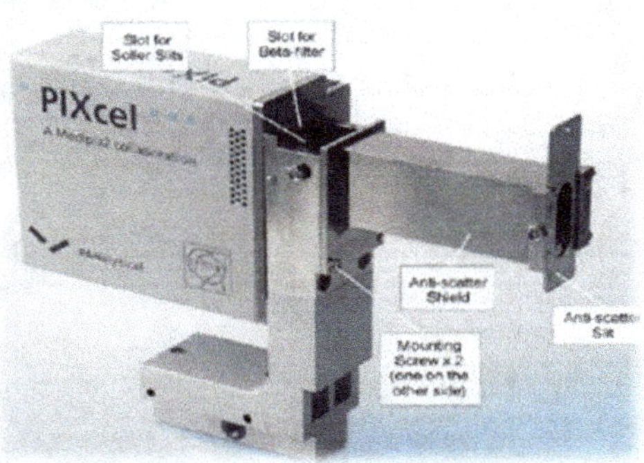

Figure 4.18 PIXcel detector [**Error! Bookmark not defined.**]

The analyses carried out by means of the diffractometer presuppose the observance of some steps that must be followed for their successful performance.

After energising the diffractometer, the cooler is put into operation and only after the diffractometer together with its computer. After accessing the diffractometer application, start the X-ray tube by means of the "BREED" button, setting the working values. Having performed these operations, the vacuum of the X-ray tube is obtained, and the anode and cathode are prepared for carrying out the analyses.

Depending on the interval between uses of the diffractometer, the procedure obtained may take between 5-40 [min]. After attaching the sample to the support, the parameters specific to the analysis to be performed, are entered into the software. The analyses include tests regarding the crystalline or amorphous state of the sample material, phase identification, chemical composition, residual stresses, thickness of small-sized layers, etc. The analysis process, depending on its complexity, can vary between a few seconds and even up to 20 hours.

By means of the detector that collects the data during the sample scan, such as the angle of reflection, the reflected beams from the sample analysis surface are processed and displayed through a graphical representation of the diffractogram type. The interpretation of the results and the final conclusions is based on the diffractograms.

The diffractometer has the possibility to support different configurations specific to the desired analyses on the sample used, as follows [**Error! Bookmark not defined.**]:

- changing the 3 supports;

- allows extraction/adjustment of the directing slots and narrowing/widening of the X-ray beam;

- allows changing the position of the X-ray cannon.

4.3 Equipment used in tribological properties research
4.3.1 CETR UMT-2 tribometer

The determination of the coefficient of friction, the hardness and the resistance to microscratches is carried out by means of the CETR UMT-2 tribometer, equipment present in the endowment of the Tribology laboratory, within the Faculty of Mechanics, the "Gheorghe Asachi" Technical University of Iaşi [**Error! Bookmark not defined.,Error! Bookmark not defined.**,118].

By means of the CETR UMT-2 tribometer, analyses of the following tribological processes can be performed, as follows [**Error! Bookmark not defined.,Error! Bookmark not defined.,Error! Bookmark not defined.**]:

- the possibility of carrying out tests regarding the determination of the friction force and the coefficient of friction (both static and dynamic), determinations by rotational and translational movement on a micro and nanometric scale, allowing tests to be carried out for a wide variety of material combinations (the standard used is ASTM 132);

- carry out tests regarding the determination of resistance to microscratches for samples with superficial deposits (the standard used is ASTM D7187);

- performs microindentation tests to determine the microhardness and Young's modulus of the material (the standard used is ASTM E18);

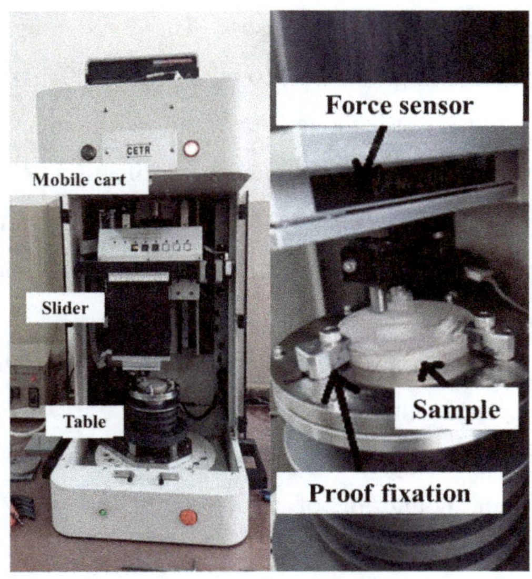

Figure 4.19 CETR UMT-2 Tribometer. Viewing the test for determining the coefficient of friction by rotational motion [**Error! Bookmark not defined.**]

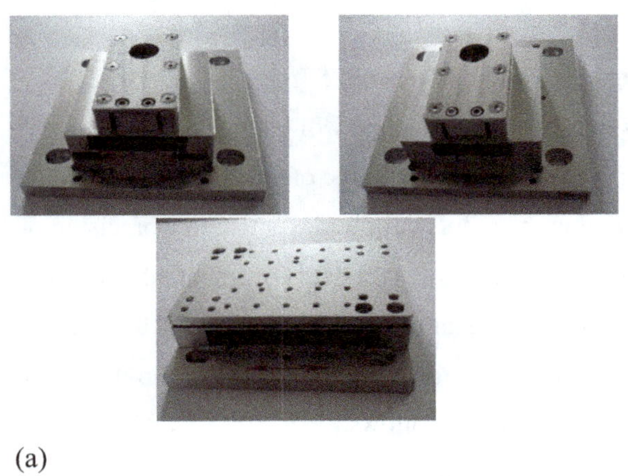

Figure 4.20 Force sensors (a) 1g; (b) 50g; (c) 2kg [**Error! Bookmark not defined.**]

The pressure and friction force measurement values are in the range of 0.1 mN and 20 N, with a resolution from 1 μN to 1 mN, depending on the measurement range of the force sensors. The force ranges covered by the tribometer are the following: 0.1 mN÷10 mN, 5 mN÷500 mN and 0.2 N÷20 N.

The system that ensures the movement of the specimen and, at the same time, the servo control of the force in the movement direction (Z direction, vertical), corresponding to the loading force of the specimen, allows the programming of the loading force (continuous, in steps, fixed), having the following characteristics, as follows:

- achieving a maximum stroke of 150 [mm];

- provides a displacement accuracy of 0.5 [μm];

- ensures a travel speed in the range of 0.002-10 [mm/s];

- allows constant monitoring of the depth of the wear mark, ensuring a precision of approximately 5 [μm].

The system that ensures the precise movement and positioning (X direction, lateral) of the test piece with servo control, has the following characteristics, as follows:

- achieving a maximum stroke of 75 [mm];

- provides a displacement accuracy of 0.25 [μm];

- ensures a travel speed in the range of 0.001-10 [mm/s];

- allows varying the radius during movement to a precision of 1 [μm];

- allows the realization of a radial movement of the pin in a spiral while maintaining constant tangential speed in the pin-disc contact.

The system that ensures the rotation of the disk through servo control, has the following characteristics, as follows:

- ensures a speed between 0.1 and 5000 [rot/min];

- ensures a maximum load of 200 [N];

- allows monitoring the number of rotations and measuring the torque;

- allows adjusting the speed to ensure a constant tangential speed in the pin-disc contact;

- equipped with a chuck for fixing discs with dimensions between 12 and 65 [mm].

To perform indentation, microscratch tests, as well as for wear processes, the equipment is equipped with the following elements, as follows:

- system that ensures pin and disc wear monitoring, with a precision of 0.25 [μm];

- equipped with a diamond tip indicator, at an angle of 120°, having a tip radius of 200 [μm];

- equipped with a linear displacement system (Y direction, horizontal), incorporated with a mass intended for fixing the sample for carrying out tests to determine the resistance to microscratches, ensuring a displacement accuracy of approximately 1 [μm], with a minimum stroke of 75 [mm] and the possibility of varying the speed, in a range between 0.001 and 10 [mm/s], achieving a displacement force of up to 1000 [N] for microscratch tests.

(a) (b)

(c)

Figure 4.21 The components of the CETR UMT-2 tribometer: (a) displacement and lateral positioning system; (b) disc rotation system; (c) horizontal displacement system [**Error! Bookmark not defined.**]

The acquisition system of the control unit that incorporates the tribometer, allows the acquisition of data of 8 independent parameters, with the possibility of reaching up to 16 parameters by using an appropriate software. Parameters can be, for example: forces; trips; temperature; time; wear; humidity, etc.

Through the use of a preset program, the CETR UMT-2 tribometer can perform tests automatically, being able to control through the software, the following parameters, as follows:

- control of the position and movements of the specimen;

- both the direction and the duration of the movement;

- force variations;

- stopping and starting the test.

The programs used for the tribometer are composed of work methods that include:

- both the working parameters and the parameters that ensure the safety of the machine;

- data regarding the test sample;

- data resulting from the test are stored in files;

- the shape of the graphics displayed on the equipment panel;

- possible calculations to be performed.

4.3.2 Instron testing machine

Static and fatigue tests are carried out by means of the INSTRON 8801 [119] testing machine, equipment present in the endowment of the Chair of Materials Resistance within the Faculty of Mechanics, "Gheorghe Asachi" Technical University in Iaşi.

The Instron testing machine [**Error! Bookmark not defined.**,120] presents the potential of performing analyses through a stability of induced quantities, namely force, displacement and deformation. A "force cell" provides force measurements with a capacity of 100 [kN]. The machine allows a travel over a length of 300 [mm] (from -150 to +150 [mm]). For carrying out measurements regarding the determination of the specific strain, the extensometer has a base of 20 or 50 [mm], achieving a maximum displacement of 30 [mm]. The hydraulic system of fixing the tank, allows changing the pressing pressure on the sample caught in the tank, so that it is possible to test some samples of materials with different hardnesses (both high-alloyed steels and brittle materials). The software (WaveMatrix) of the machine ensures the successful completion of the tests, but for each test, it is necessary to prepare the test program with the specific data in advance. Tests can become predictable through the performance that characterises the test machine.

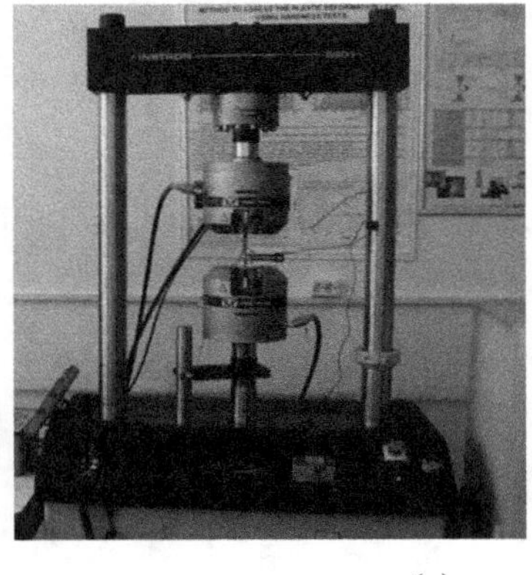

(a)

(b)

Figure 4.22 The Instron 8801 testing machine (a), together with the computer that equips it (b) [**Error! Bookmark not defined.**]

This testing machine can perform both direct tensile and compression tests, as well as bending and shear tests, by means of the specific devices in its equipment.

By means of specific programs designed to perform testing on the machine, it can perform the following tests as follows, but not limited to these:

- tests until the final break;

- tests with leveling of force (displacement or specific deformation);

- repeated attempts.

The work methods that make up the test program, provide for the following:

- working and safety parameters of the machine;

- data regarding the sample subjected to the test;

- parameters specific to the test;

- result data (stored in files);

- the shape of the graphics (from the car's computer screen);

- possible calculations to be made.

These programs allow the determination of all parameters, both standardised and non-standard parameters.

To determine the fatigue stresses, it is necessary to carry out some test methods that allow both the control of the machine and the stability of the test parameters.

By performing fatigue tests, it is possible to determine:

- parameters for starting loading cycles (average force, relative position of the specimen, etc.);

- minimum and maximum force;

- the frequency of the request;

- cycle shape (sinusoidal; triangular and rectangular/square);

- the number of cycles of the request (in the case of tests until the final break).

4.4 Methods and equipment used in performing electrochemical analyses
Potentiostat equipment for the corrosion study

A method for characterising corrosion processes, is cyclic potentiometry. In this method, the polarisation curves are drawn, successively for first increasing and then decreasing values of the potential of the working electrode, recording the current in the circuit. There are several ways of representing the data, depending on the purpose pursued and the process to be highlighted. More often used are the representations $J = f(E)$ or $I = f(E)$ and the semi-logarithmic representation: $E = f(\log J)$. Sometimes, these curves are improperly called "hysteresis curves" because of their similar shape [121,122].

The equipment intended for carrying out the electro-corrosion tests, together with the working premises, are presented in Figure 4.23.

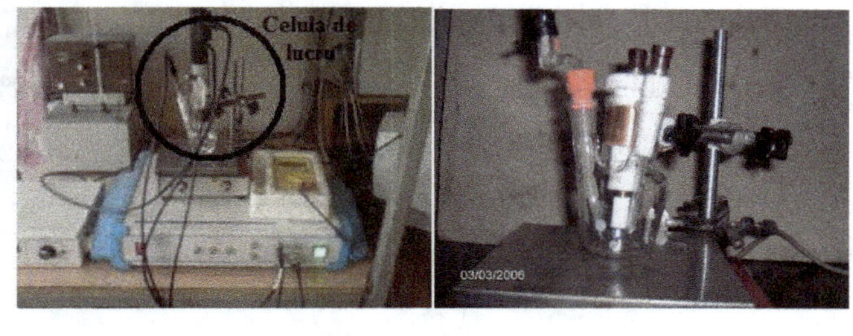

(a)   (b)

Figure 4.23 Corrosion investigation equipment on metallic materials: (a) potentiostat equipment and (b) work cell

From the analysis of cyclic polarisation curves (cyclic voltammograms), information can be obtained regarding the type of electrochemical process that takes place at the electrode/medium interface (such as: generalised corrosion, localised corrosion, passivation, reductions or oxidations of the species in the solution), the evaluation characteristic potentials (corrosion potential, penetration potential, repassivation potential, transpassivation potential, protection potential) and the signaling of secondary electrochemical processes that take place, either on the surface of the electrode or in the mass of electrolyte (oxidations, catalysed processes, etc.).

Corrosion behavior is followed by rapid electrochemical tests and by dynamic potentiometry. Open-circuit potential measurements and dynamic polarizations are performed using a Volta Lab 21 potentiostat (Radiometer, Copenhagen), the equipment's operating diagram is shown in Figure 4.24 [**Error! Bookmark not defined.,Error! Bookmark not defined.**,123].

To calculate the corrosion rate of an alloy immersed in a corrosive environment, it is necessary to know the instantaneous current density, which is determined by the polarised resistance method. This method is used to determine the corrosion current at the corrosion potential of a metal or alloy, using for this purpose, the linear polarisation curve obtained for relatively small overvoltages. The corrosion current determined in this way represents the corrosion current that occurs at the metal-medium interface when the metal is immersed in the solution and can be measured directly by electrochemical methods.

The principle diagram of an electrochemical corrosion investigation equipment is presented in Figure 4.24 and consists of the following elements: 1- electrochemical cell, 2- auxiliary electrode; 3- amplifier, 4 – working electrode, 5- reference electrode, 6- reference source, 7- multiturn potentiometer, 8,9- power supply.

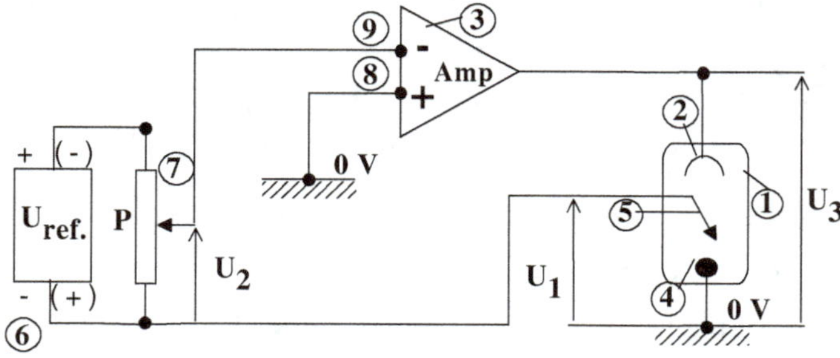

Figure 4.24 Schematic diagram of the electrochemical corrosion investigation equipment

Data acquisition and processing is done with the Volta Master 4 software. A three-electrode cell equipped with a stirring system is used, with the specification that the cylindrical electrodes are mounted in Teflon supports that allow connection to the electrochemical system's rotation electrode.

Electrochemical tests (potentiodynamic polarization, EIS - electrochemical impedance spectrometry) are used to detect pores and microholes in the deposited layers, as well as their effects on the corrosion resistance behavior, for a short immersion period. Figure 4.25 shows the image of the potentiostat used for corrosion resistance analyses.

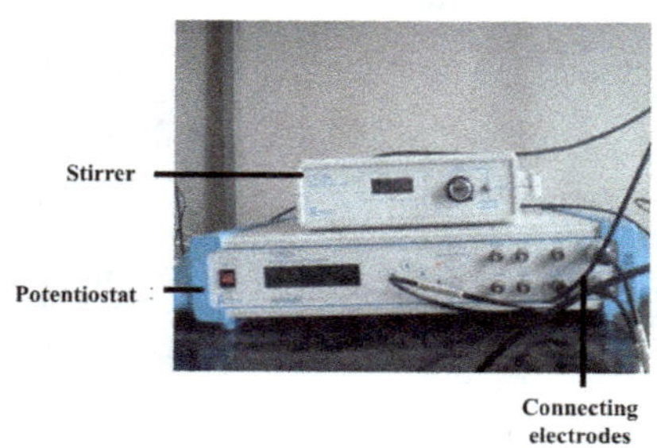

Figure 4.25 Potentiostat equipment used for corrosion study

EIS measurements provide important information regarding the surface layers of coated samples.

General aspects regarding electrocorrosion of metallic materials

The method used to test the resistance to corrosion and electrocorrosion, is based on the evaluation of the polarization resistance, R_p, which is defined as the slope of the tangent to the potential-current density curve [E = f(j)], at the equilibrium point (E = E_0 sau η = 0), i.e. at the free corrosion potential:

$$R_p = \left[\frac{\Delta E}{\Delta j}\right]_{E=E_o}$$

(5.1)

From a theoretical point of view, the basis of the method is the Butler-Volmer equation, which, if metal corrosion occurs at the electrode, through a single electrode reaction controlled only by charge transfer, can be written in the form:

$$j = j_{cor}\left[\exp\left(\frac{2{,}303(E - E_{cor})}{b_a}\right) - \exp\left(-\frac{2{,}303(E - E_{cor})}{b_c}\right)\right]$$

(5.2)

where: b_a and b_c are the Tafel slopes for the two branches, anodic and cathodic:

$$b_a = \frac{RT}{\alpha z F} \quad \text{respectiv,} \quad b_c = \frac{RT}{(1-\alpha)z F}$$

(5.3)

$(E-E_{cor})$ - is the overpotential of the electrode, and Jcor – the corrosion current density at the corrosion potential E_{cor} (or at zero overpotential).

It has been observed experimentally that J varies approximately linearly with the applied potential (E), starting from approximately 50...60 mV versus the corrosion potential and only in a range of approximately 10...20 mV [3].

Researchers Stern and Geary simplified the Butler-Volmer equation (eq. 2) for the case of small overpotentials compared to E_{cor}, mathematically linearising this equation by developing the exponential terms in series ($e^x = 1 + x + x^2/2!$ $+ x^3/3! +$ and neglecting higher order terms.

The obtained simplified equation can be transcribed in the form:

$$R_p = \left(\frac{dE}{dj}\right)_{(E_{cor})} = \frac{b_a \cdot b_c}{2{,}303 \cdot J_{cor} \cdot (b_a + b_c)} \; ; \quad (ohm.cm^2)$$

(5.4)

After rearranging the equation, it can be obtained for the instantaneous corrosion current:

$$J_{corr} = \frac{b_a \cdot b_c}{2.303(b_a + b_c) \cdot R_p'} \quad (mA/cm^2)$$

(5.5)

Polarization resistance can be expressed in absolute value or relative to the surface unit depending on the quantity considered on the current axis (current intensity or density), the units of measure for resistance depending on the units of measure for E and I or j. If E is expressed in mV and I in mA, or E in V and I in A, then Rp is expressed in ohms.

$$R_p = \left(\frac{dE(mV)}{dI(mA)}\right)_{E_{cor}} = tg\alpha \quad (ohm)$$

(5.6)

respectively,

$$R_p' = \left(\frac{dE(mV)}{dJ(mA/cm^2)}\right)_{E_{cor}} = tg\beta \quad (ohm.cm^2)]$$

(5.7)

To calculate the instantaneous corrosion current, the Tafel constants, b_a and b_c, are required, which are the slopes of the linear portions of the anodic and cathodic branches of the polarisation curve in coordinates $E = f(\log j)$ sau $E = f(\log I)$:

$$b_a = \left(\frac{dE}{d\log J}\right)_{anodic} \text{(mV)} \quad \text{respectively} \quad b_c = \left(\frac{dE}{d\log J}\right)_{catodic}$$

(mV)   (5.8)

and the polarisation resistance, Rp - as the slope of the tangent to the polarisation curve, the zero overpotential point (at the point E=Ecorr).

The corrosion current density is obtained in mA/cm2 if R' is expressed in ohm.cm2, and b_a and b_c in mV/decade or in A/m^2, if R' is expressed in ohm.m^2 and the Tafel constants in V/decade.

The corrosion rate is calculated with the relation:

$$v_p = 3{,}27.\overline{\left(\frac{A}{z}\right)}.\frac{J_{cor}}{\rho} \qquad \text{[mm/an]}$$

(5.9)

where **A** - represents the atomic mass of the chemical element in the alloy that undergoes electrochemical corrosion, **z** - is the electric charge of the metal ion passed into the solution, ρ - the density of the alloy and **J** - the density of the instantaneous corrosion current.

The corrosion current determined in this way actually represents the corrosion current that occurs at the metal/corrosive medium interface when the metal is immersed in the solution and can not be measured directly by electrochemical methods. This is actually an instantaneous corrosion current.

CHAPTER 5 Experimental results of obtaining superficial deposits by the Cold Spray method

5.1 Planning the conduct of experimental research

Planning to carry out experimental research involves going through the following stages, as follows:

(a) Microstructural analyses:

- Optical microscopy;

- SEM electron microscopy.

(b) Determination of corrosion resistance:

- Electrochemical corrosion processes by potentiodynamic testing methods and electrochemical impedance spectroscopy.

(c) Analysis of mechanical properties:

- Determination of the coefficient of friction through rotational and translational movement;

- Microindentation;

- Microscratches;

- Traction;

- Fatigue.

5.2 Deposits made by the Cold Spray method

Surface depositions were performed using the VRC Gen III Cold Spray (VRC Metal Systems). The depositions on the used samples were carried out through the international collaboration with Northeastern University in Boston (USA) finalising a research contract "Program in materials, Additive and Secure Manufacturing and Multiscale Materials Engineering (3ME)", Grant number: W911NF2020024. The project is carried out within the US Army Research Laboratory (DEVCOM), in collaboration with several universities and research institutes: MIT, MMT, CTC, UCONN, VRC Metal Systems and the "Gheorghe Asachi" Technical University of Iasi (TUIASI) . This collaboration gives the opportunity to contribute to the consolidation of information regarding the thermal spraying method "Cold spray" and, at the same time, access to several samples deposited by this method on materials used in military applications.

The powders used in covering the samples are based on chromium and nickel carbide, and this type of powder is called WIP-C1. Previously, in order to provide a much stronger bond between the deposited layer and the base substrate, a powder deposition was applied exclusively for the creation of a bonding layer called WIP-BC1. This bonding layer also exhibits properties to improve the adhesion of the layer with WIP-C1.

Cylindrical specimens were fixed on a lathe, and coating with WIP-C1 was accomplished by axially moving the coating nozzle while keeping it normal to the axis of the specimen. The parameters used for spraying the powder mixture are specified in Table 5.1.

Table 5.1 Working parameters for cold deposition of samples

Parameter	Value
Gas	Nitrogen
Pressure	6.2 MPa (900 psi)
Temperature	675°C
The ID of the nozzle	WC NZL0060
Nozzle neck size	2 mm
Powder feed rate	10 rpm
Powder feed gas flow rate	105 slm
Spray distance	25 mm
Spray angle	90 deg.
The passing speed of the nozzle	250 mm/s
Nozzle pitch distance	0.25 mm
Layer thickness	0.127 mm
Target coating thickness	0.508 mm
Powder type	WIP-C1
Substrate powder type	WIP-BC1 at an angle of 60°

To perform a deposition, the system is pressurised to the reference value, namely 900 [PSI] for these depositions.

At this point, the gas begins to heat up, and this process takes about 10 minutes.

As soon as the temperature reference value is reached, the powder feeder is turned on, and by means of the robotic arm, the nozzle is directed to carry out the coatings at supersonic speeds, resulting in the mechanical interlocking of the particles and thus leading to the formation of the superficial coating layer. Figure 5.1 shows the samples during the depositions, and in Figure 5.2, the samples with depositions after the tests.

(a)

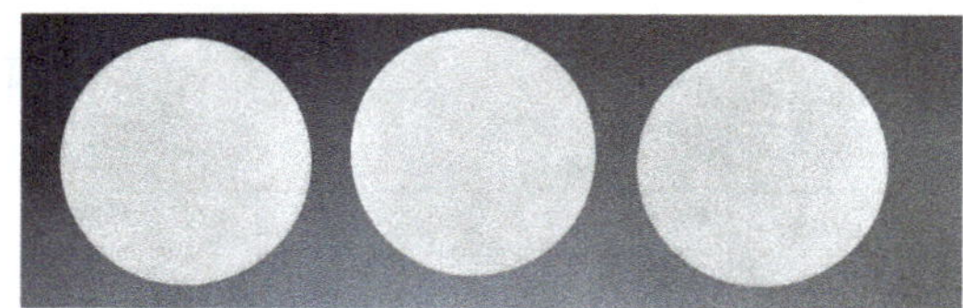

(b)

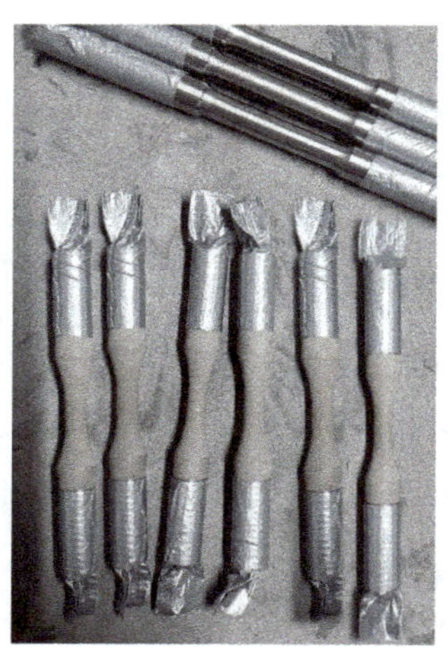

(c)

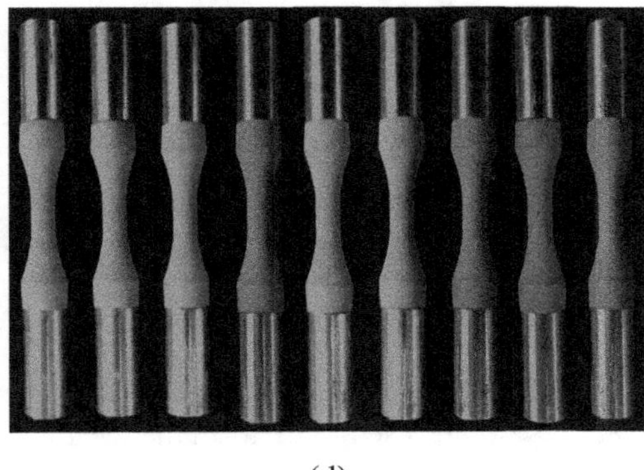

Figure 5.1 Applicator and nozzle mounted on robotic arm for cold deposition (a) and samples after deposition (b-d)

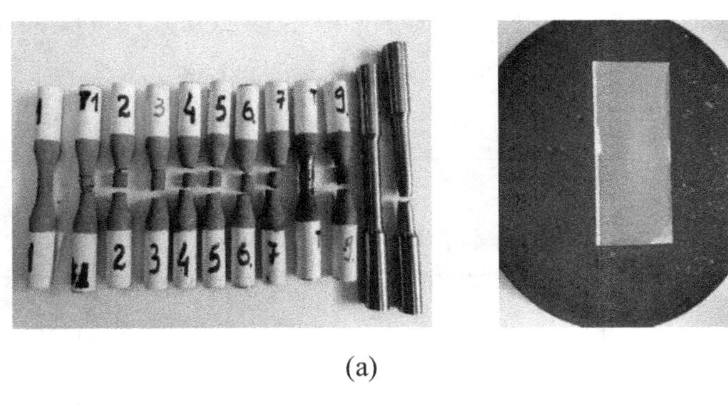

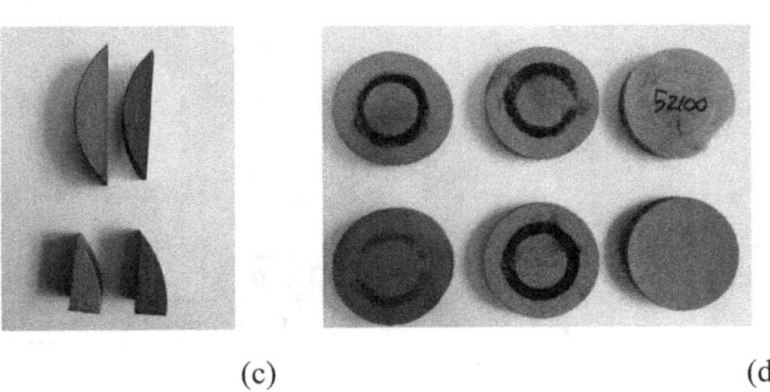

(e)

Figure 5.2 The samples used to perform the tensile analyses (a); microstructural (b); microscratch (c); determination of friction coefficient by rotational movement and microindentation tests (d); determination of the coefficient of friction by translational motion (e)

The dimensions of the samples used in the tests are presented in Table 5.2.

Table 5.2 Samples dimensions

Sample	Dimensions	Destination
Round test tube	88 x Ø6 [L x D]	Tensile test
Rectangular parallelepiped type specimen	23 x 10 x 10 [L x l x h]	Optical microscopy
Rectangular parallelepiped type specimen	64 x 20 x 10 [L x l x h]	Determination of apparent friction coefficient by translational motion
Rectangular parallelepiped type specimen	20 x 10 x 10 [L x l x h]	Microscratch tests
Disc type specimen	Ø64 x 10 [D x h]	Determination of apparent friction coefficient by rotational motion
Disc type specimen	Ø12 x 3 [D x h]	Corrosion tests
L=length; l=width; h=height; D=diameter		

5.3 Structural analysis of the surface layers deposited using WIP-C1 powders by the Cold Spray method by means of optical and electron microscopy

Chromium carbide and Ni agglomerated particles were deposited on AISI 52100 steel. Figure 5.3 (a,c) shows the surface of the base material at different magnifications, and Figure 5.4 (a,e) illustrates the deposition morphology in cross section. In Figure 5.4, it can be seen that the particles entered the base substrate, simultaneously deforming following the collision both the particles and the substrate which has a high hardness which increases the degree of deformation of the particles and which, at the same time, allowed the deposition of the particles resulting in obtaining of dense coatings and closely related to the surface of the substrate. In Figure 5.4 (a), it can be seen that the deposited surface is flat, without cracks in the coating material or in the particles in the coating material.

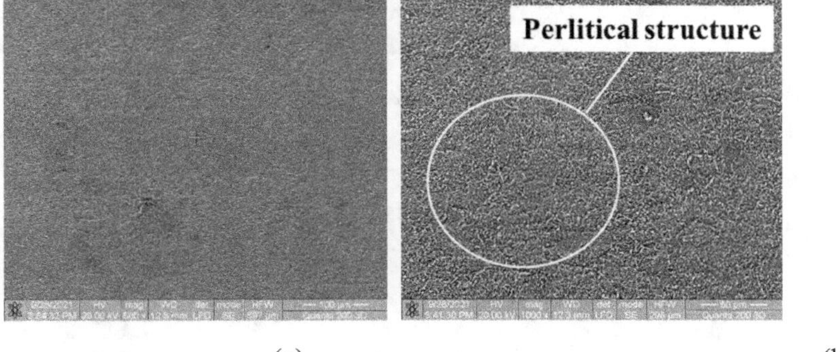

(a)   (b)

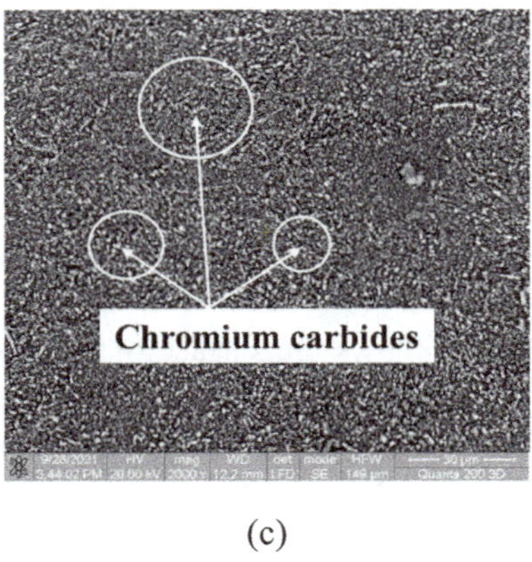

(c)

Figure 5.3 Base material micrograph (a) 500X; (b) 1000X; (c) 2000X

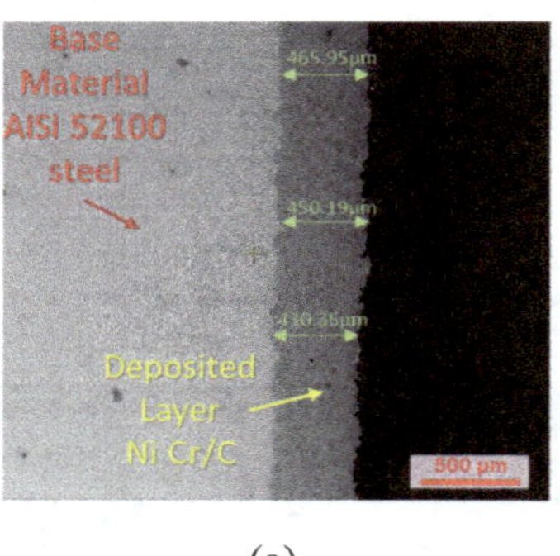

(a)

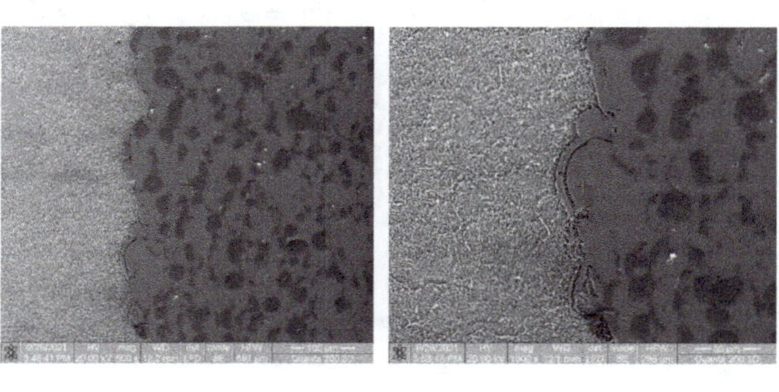

(b)

(c)

96

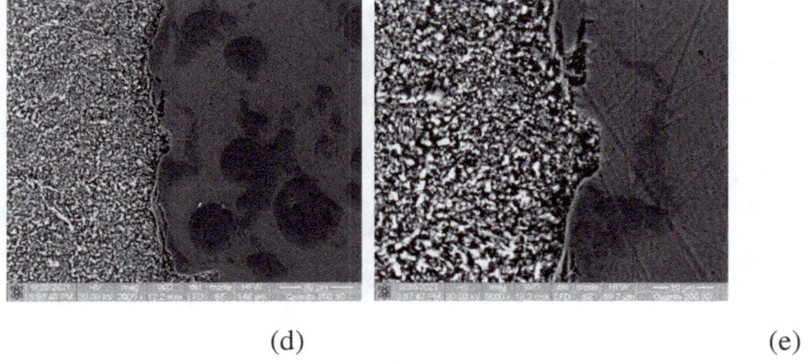

(d)   (e)

Figure 5.4 Cross-sectional view micrograph of the coated material (a) 100X; (b)500X; (c) 1000X (d) 2000X și (e) 5000X

As it can be seen from the morphological appearance of the particles in Figure 5.5, the particles are of different sizes, thus allowing the particles to penetrate and deform the substrate where the small-sized particles are propelled at a higher speed, thus leading to deeper penetration into the surface substrate. The particle size is between approximately 17 and 66 µm (Figure 5.5 b), and according to EDS tests, the chemical composition was composed of Ni, Cr, C and O (Figure 5.5 f). Table 5.3 shows the mass percentage of the component elements of the deposited layer, after EDS tests.

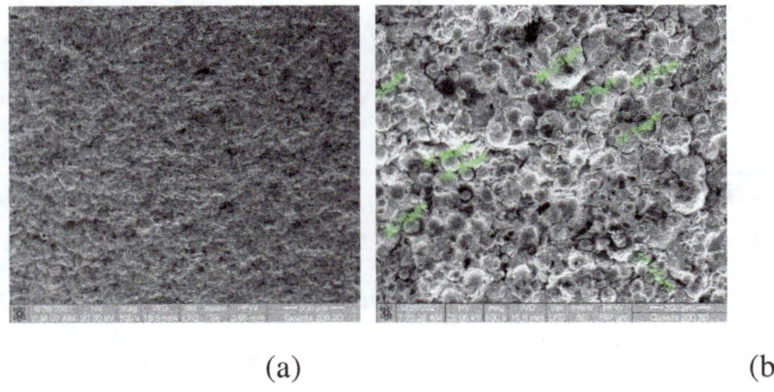

(a)   (b)

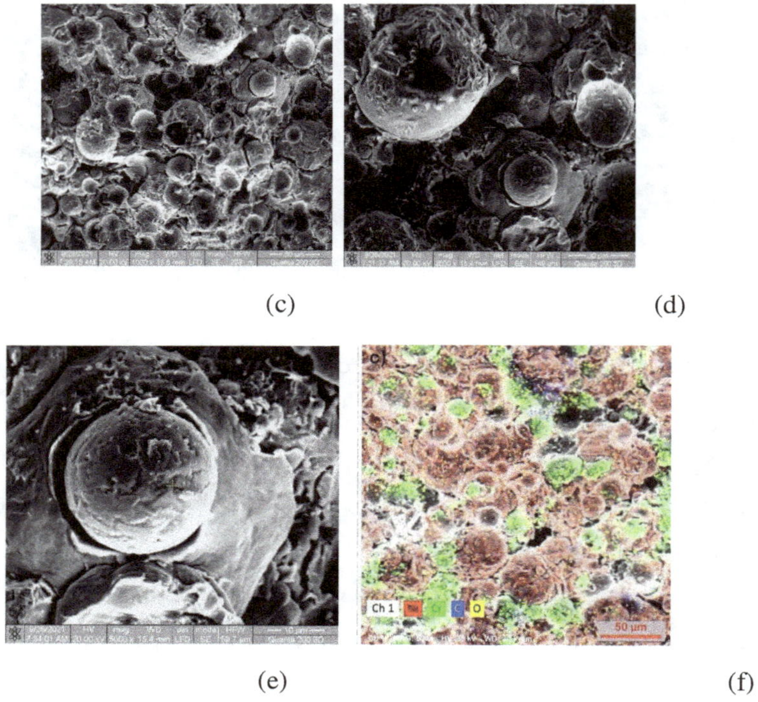

(c)   (d)

(e)   (f)

Figure 5.5 Surface morphology of superficial deposits (a) 100X; (b) 500X; (c) 1000X; (d) 2000X; (e) 5000X and (f) EDS mapping surface

Table 5.3 Chemical composition of WIP-C1 powder coating determined by EDS tests

Composition (%)	Cr (%)	O (%)	Ni (%)	C (%)
WIP-C1	22.25	3.44	57.65	16.64

The structural analysis of the superficial layers was carried out by X-ray diffraction, so that in Figure 5.6, the diffractograms generated for both the base and the coating material, are presented [124]. From the diffractograms, one can see the predominant phase of NiCr (cubic structure/ICDD 96-901-2971), which was obtained at a 2 Theta angle value of about 44°, the secondary phase of Cr_3C_2 (cubic structure/ICDD 96-901-1599) at a 2 Theta angle value of 53° and

98

77°. From what can be seen in the base material, the predominant phase is Fe (cubic structure/ICDD 96-901-6602), at an angle of approximately 2 Theta of 45°, the secondary phases being Fe and Cr_3C_2 at a value of the Theta angle of 65° and 83°.

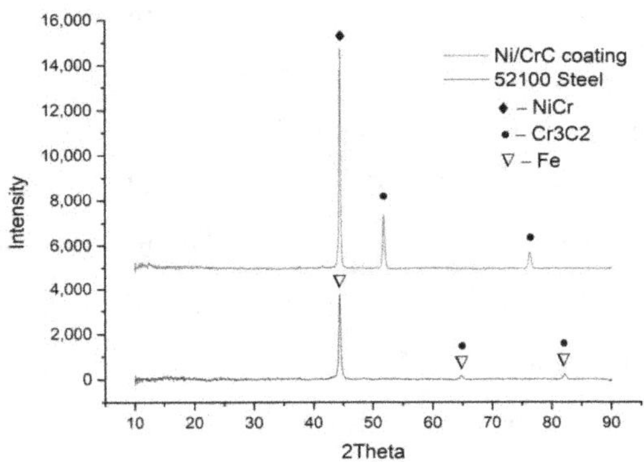

Figure 5.6 X-ray diffraction for both base and cover material [**Error! Bookmark not defined.**]

5.4. Partial conclusions

1. At first glance, Ni/CrC based powders form a very stable coating on 52100 alloy steel when deposited by cold spraying.

2. Microstructural analysis shows that the deposited layers are dense and closely connected to the surface of the substrate, and through the simultaneous collision and deformation between the particles and the underlying substrate, the particles are deeply embedded, realising this strong connection through a mechanical interlocking phenomenon.

3. From X-ray diffraction, it can be seen that, for the base material, the predominant phase is Fe, and for the layer deposited by Cold Spray, with WIP-C1 powders, the predominant phase is NiCr.

CHAPTER 6 Experimental results on the physical-mechanical characteristics of the superficial deposits made by the Cold Spray method

6.1 Determination of fatigue behaviour

6.1.1 Static tensile tests

Before the samples are subjected to fatigue loading, certain characteristics of the material from which these samples are made, must be determined. These characteristics are determined by testing the samples, first under static tensile stress. It is necessary to determine, first of all, the yield strength of the material, from which one starts, with lower values, in the fatigue stress.

Static tensile stress tests are performed in accordance with ASTM E8/E8M-16 (Standard Test Methods for Tension Testing of Metallic Materials).

The tensile tests were performed directly on a coated sample, to be sure that the characteristics determined in static traction are those of the material that will be subjected to cyclic fatigue loading. Under these conditions, the covered sample, of the type and shape that can be seen in Figure 6.1 (a), was requested for static traction and mounting it on the Instron test machine (b) [**Error! Bookmark not defined.**].

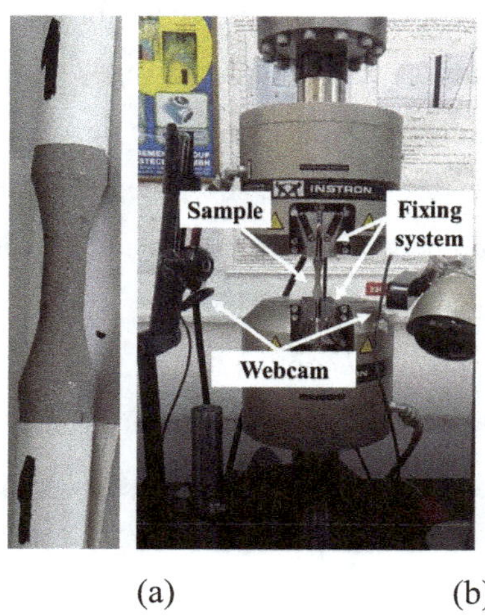

(a)   (b)

Figure 6.1 Sample with deposits subjected to tensile testing (a) and mounting on the Instron machine (b)

A strain gauge was mounted to be able to obtain better accuracy in acquiring specific strains. As a result, the specific strains in Figure 6.2 are those taken from the strain gauge. Since the total deformation of the sample was not very large, strain gauge data could be taken until the sample broke.

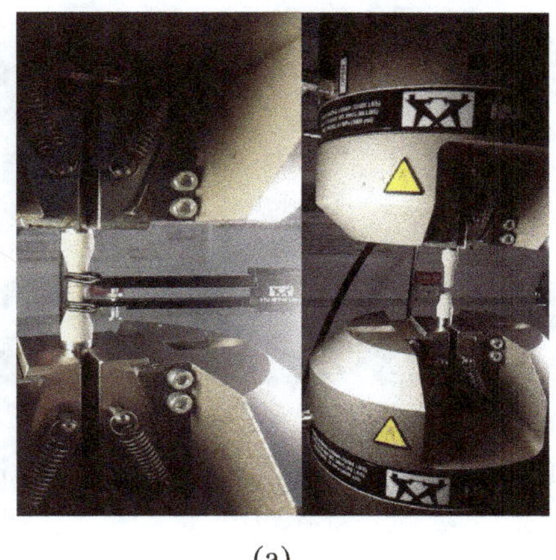

(a)   (b)

Figure 6.2 Strain gauge attachment for static tensile testing

The broken sample, with base material 52100, can be seen in Figure 6.3. Both from the examination of the broken sample, Figure 6.3 (a), and from the characteristic curve in Figure 6.3 (b), the following conclusions can be drawn, after the static tensile stress of a coated sample:

- First, we mention the fact that the shape of the characteristic curve is given, to the greatest extent, by the behavior of the base material, in this case, it was material 52100;

- The initial zone of elasticity is a large one, in which high values of tension are reached on a linear portion;

- There is no area of cold-hardening of the material; after reaching the yield strength, Rp0.2 or (σp0.2), the material undergoing pronounced elongation, but no further increase in strength;

- Given that the total elongation is 15%, finally, when breaking is reached, the base material undergoes significant plastic deformations, hence the necking observed in Figure 6.3;

- Also due to the mentioned total elongation, the deposited layer suffers significant destruction.

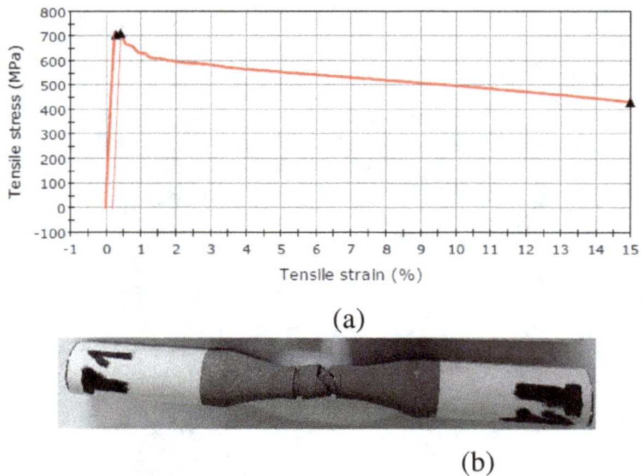

(a)

(b)

Figure 6.3 Specific stress-strain curve (a) and broken sample after static tensile stress (b)

It is found that the stress σp0.2 is 831.6 MPa, a value from which we start with decreasing values in the fatigue stress.

Table 6.1 The characteristics obtained after the static tensile stress of the coated sample

	Modulus (Automatic Young's) (MPa)	Tensile strain at Yield (Offset 0.2 %) (%)	Tensile stress at Yield (Offset 0.2 %) (%)	Extension at Yield (Offset 0.2 %) (mm)
1	-	1.81796	831.60462	0.29087
	Load at Yield (Offset 0.2 %) (N)	**Tensile strain at Tensile Strength (mm/mm)**	**Tensile stress at Tensile Strength (MPa)**	**Extension at Tensile Strength (mm)**
1	23513.06610	0.03664	682.17316	0.58623
	Load at Tensile Strength (N)	Tensile strain at Break (Standard) (mm/mm)	Tensile stress at Break (Standard) (MPa)	Extension at Break (Standard) (mm)
1	19287.99152	0.13838	503.74460	2.21404
	Load at Break (Standard) (N)	**Energy at Break (Standard) (J)**	**Specimen note 1**	**Geometry**

1	14243.04247	37.24281	Covered break sample 52100	Circular
	Length (mm)	**Thickness (mm)**	**Width (mm)**	**Diameter (mm)**
1	16.00000	-	-	6.000000

6.1.2 General considerations on fatigue testing

Fatigue tests were performed in accordance with ASTM E466-15 standard (Standard Practice for Conducting Force Controlled Constant Amplitude Axial Fatigue Tests of Metallic Materials), and the dimensions of the fatigue test specimens are shown in Figure 6.4.

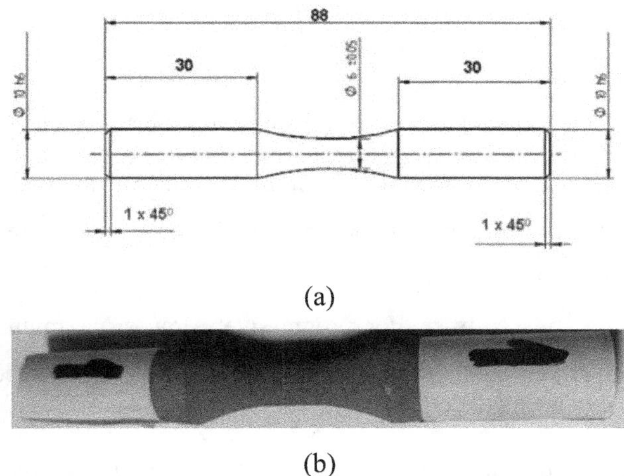

(a)

(b)

Figure 6.4 Dimensions of fatigue test specimens (a); shape and appearance of coated samples (b)

The fatigue tests were carried out following a symmetrical alternating cycle (Figure 6.5), having the following characteristics:

- The maximum voltage value is equal to the minimum voltage value (in absolute value);

- The average voltage, σm, has zero value;

- Cycle asymmetry factor R=σmin/σmax=-1;

- The amplitude of the stress cycle has the value of the maximum voltage.

As a result, the samples were tested in both tension and compression with the same force (stress) value. This was possible as a result of the facility offered by the testing machine, equipped with jaws through which the sample is hydraulically tightened.

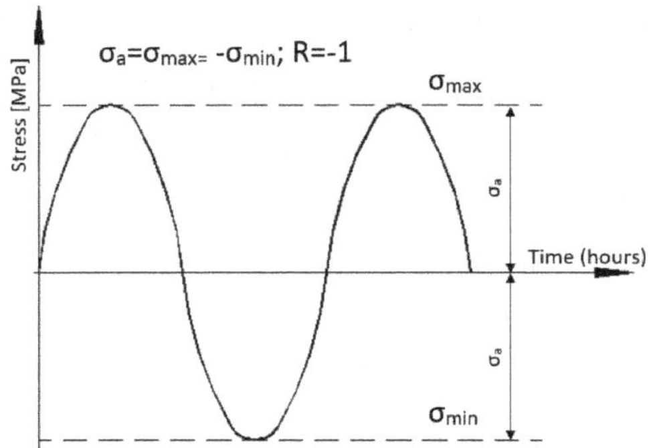

Figure 6.5 Form of the fatigue cycle used in the tests

It is mentioned that, for the fatigue stress, the values of the initial, test stresses must be in the elastic range. The fatigue test is performed on several samples, starting with a stress value close to the yield strength and continuing by decreasing the stress further. For progressively lower maximum test stresses, progressively higher numbers of cycles to failure will be obtained. As a result, the first maximum test stress was at 90% of the yield strength value, taking into account the fact that the elastic region reaches close to the maximum stress, respectively:

$$\sigma_{max} = 0.9 \cdot \sigma_{p0.2} = 0.9 \cdot 710 = 636 \ \text{MPa}$$

(6.1)

In Figure 6.6, you can see the screen of the fatigue test program, after a sinusoidal, symmetrical alternating cycle. Here, the loading force was ±12.8 kN, corresponding to a stress of approximately 454 MPa. In the upper left part of the image, the variation of the position of the test piston is given, in the upper right part of the image, the approximate form of the cycle is shown (points taken at different moments of the test), in the lower left part of the image, the variation of the force test, and the lower right part of the image shows the maximum and minimum force during the fatigue tests.

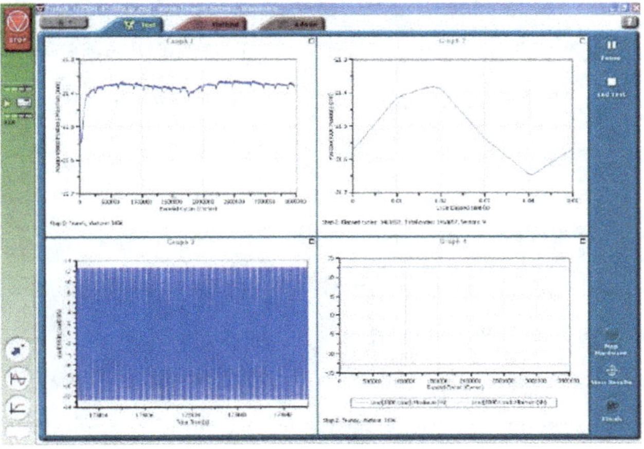

Figure 6.6 Interface of the fatigue test program

We note that the main objective was to observe the behaviour of the coating material under fatigue stress and not to plot the fatigue life diagram for the base material. It is true that there must be a compatibility between the deformation experienced during cyclic fatigue stress by the two materials; the base material and the covering material. If this compatibility did not exist, major separations of the coating material from the base material would occur. These separations are materialised by the appearance of cracks at the interface between the two materials, separation of the coating material also at the interface, the appearance of intra- and/or intergranular cracks in the coating material, along the front of the propagated crack. As a result, in the macroscopic and microscopic observations below, the failure/destruction aspects, especially regarding the coating material, in different areas of the fractured surfaces will be followed in particular.

It is known that the two materials, the base and the coating, at the same applied voltage will have a different specific strain, respectively $\varepsilon_{base\_material} > \varepsilon_{coated\_material}$. In the conditions in which the two materials will deform differently, at the same moment of time, respectively, the base material will deform more, there is a possibility that at least within the large deformations that occur at higher values of the stress cycles, significant damage to the covering material occurs. To be able to observe this, images were taken during the fatigue test, from the central, calibrated area of the samples. As it can be seen from Figure 6.7, and from all the pictures taken and studied, in none of the cases, during the tests, no damage to the covering material was observed. This was not observed even at a large number (over a million) of the request cycles. As it will be seen later, from the SEM images, the adhesion of the coating material to the base material, in all cases, was good, with no significant differences in the fatigue behaviour of the coating material and the base material, which lead to damage first to the coating material and then to the base material as we would expect. As a result, a macroscopic analysis shows a very good behaviour of the covering material under symmetrical alternating tensile stress, the stress occurring both in tension and compression.

Figure 6.7 Appearance of the outer surface of the samples

In Figure 6.8, the moment of fatigue crack propagation was captured. It is known that, depending on the material, a relatively large number of fatigue cycles may be required for crack initiation. After crack initiation, depending on the magnitude of the stress, the number of fatigue crack propagation cycles can also be higher or lower. In Figure 6.8, the crack that can be seen (at the bottom and with the approximate shape of the artificially drawn line) is the one propagated by fatigue, a relatively small number of cycles being required before the final, sudden rupture. A remarkable fact can be seen from this figure, namely that although we have a macroscopic crack propagation and we are close to the final fracture, no material separation is observed, from the material used to cover the sample, near the front of the propagated crack. As a result, the crack propagation in the cover material follows the same path as in the base material, without larger collateral damage occurring in the cover material. In these conditions, at least from a macroscopic point of view, we can say that we have a very good behaviour of the covering material, when subjected to cyclic fatigue, after a symmetrical alternating cycle.

Note: Symmetric alternating stress is known to be the least favorable of cyclic fatigue tests.

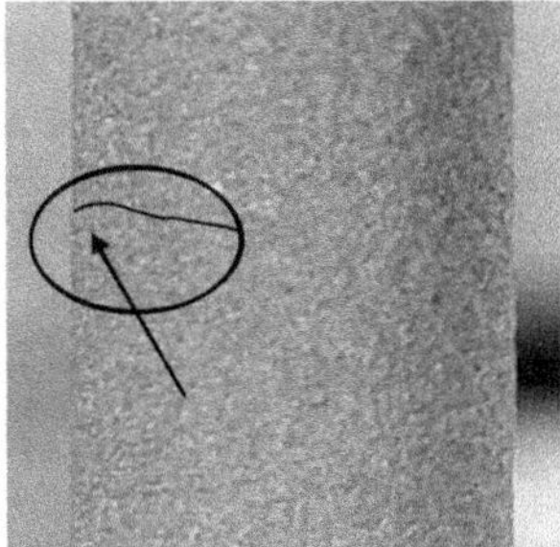

Figure 6.8 Fatigue crack propagation

As previously mentioned, the tests were conducted from higher values of stress (resulting in lower values of the number of stress cycles) to lower values of stress. The value of the number of request cycles of 5 million was set as a reference. If up to this value of the number of stress cycles, regardless of the stress tension, the specimen did not break, the test was stopped. Considering that, depending on the applied voltage, the request time could be long, to follow the behaviour of the samples during the requests, two web cameras were mounted with the help of which the test could also be followed remotely, on the mobile phone. In Figure 6.9 (b), a fatigue fractured specimen is shown, here being a high stress and a small number of stress cycles.

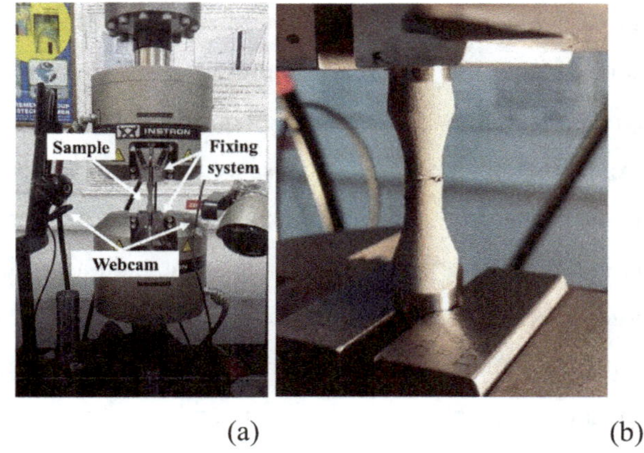

(a)   (b)

Figure 6.9 Tracking of the two-web camera tests (a) and the fatigue-broken sample (b)

6.1.3 Fatigue determinations performed on coated samples

Preliminary considerations regarding the performance of fatigue tests:

For the fatigue test, 9 samples were used, made of 52100 steel, Figure 6.10. Samples 1 and 8 did not break even after 5 million cycles, therefore the test was stopped and the samples were kept unbroken. The other samples broke at different stresses and cycle numbers.

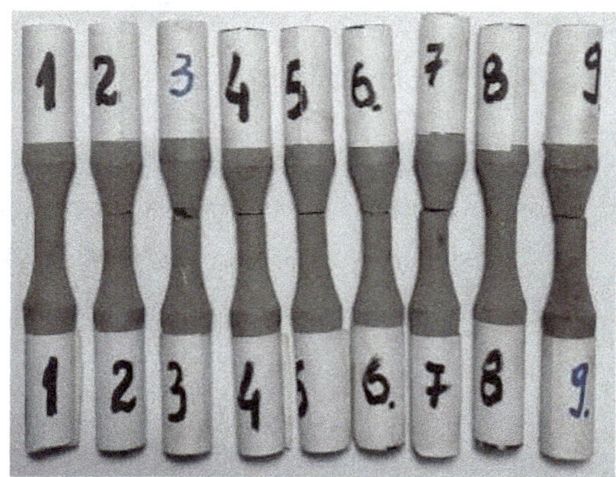

Figure 6.10 Viewing samples after testing

The macroscopic appearance of the surfaces was observed for both unbroken and broken samples. As it can be seen in Figure 6.11, no detachments of the deposited material resulted even after a large number of cycles. It is true that the test stress in this case was a small one, 420 MPa, relative to the yield strength of 852 MPa. At low stresses, in the elastic range, the deformations of the base material do not differ much from the deformations of the coating material. Under these conditions, during loading, the expansion and compression of the cover material follows, to the greatest extent, the expansion and compression of the base material. This is the reason why, although the demand is made at a high number of cycles, over 5 million, the stress being low, the coating material resists very well under cyclic fatigue loads, without significant damage being observed.

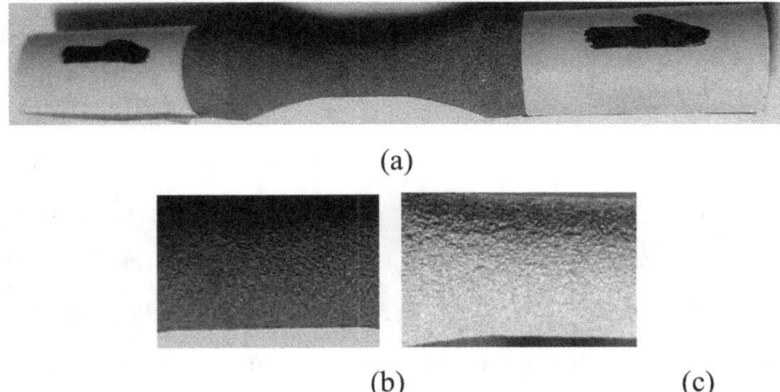

(a)

(b)   (c)

Figure 6.11 Macroscopic appearance of sample 1 after fatigue stress (Stress: σ = 420 MPa, Number of cycles: N = 5519600)

For each sample subjected to alternating-symmetric fatigue stress, 4 macroscopic photographs were taken, as follows:

- side view of the broken sample after the fatigue test, Figure 6._a:

- side view with the approach of both ends resulting from the break, Figure 6._b;

- image with both ends resulting from tearing, Figure 6._c, frontal view, made immediately after the request;

- photo of one of the surfaces resulting from the break, Figure 6._d, front view, taken before analysis in the SEM Quanta 3D electron microscope.

With the SEM Quanta 3D microscope, 5 photographs were taken, which joined the macroscopic ones, for the identification and observation of the breaking parameters. These were:

- an overall photo, with the magnitude of 42X, to be compared with the photo taken on the electron microscope, Figure 6._e;

- a photo of the final, sudden break area, Figure 6._f;

- two close-up photos in the crack initiation region with magnitudes of 100x and 200x, Figures 6._g and 6._h;

- a photograph with the magnitude of 500x of the nucleation/initiation zone of the defect with the highlighting of the initial crack, Figure 6._i;

Sample no. 2. Tension = 636 MPa. Number of cycles = 5300

Macroscopic observations

In the area of crack initiation, the surface of the coating material is flat, without obvious macroscopic cracks. Throughout the crack propagation, the surface of the covering material remains flat, with no material separation, Figure 6.12 (c). At the boundary between the areas of the slowly propagated fatigue crack and the suddenly propagated crack, significant damage to the coating material is observed, Figure 6.12, (d), (e) and (f) is observed.

Microscopic observations

In the area of crack initiation, no significant damage to the covering material layer is found, Figure 6.12 (g), such as macroscopic detachments of the material, cracks inside the covering material, cracks of the particles that constitute the covering material. In contrast, in this area of crack initiation, Figure 6.12 (g) and (h), damage can be seen in the area of the coating material with microscopic material detachments. Considering that, in this area, on the outside of the sample, no damage to the covering material can be found, it can be concluded that the crack first started in the base material, after which it extended in both directions, both in the covering material, to the outside, as well as in the base material, to the inside. In Figure 6.12 (i), a detachment of the covering material from the base material is observed. This detachment is observed only in this area, which leads us to the conclusion that the crack initiated here. Detachment of the covering material from the base material occurred as a result of the differences between the elastic moduli of the two materials, thus the differences in specific deformation.

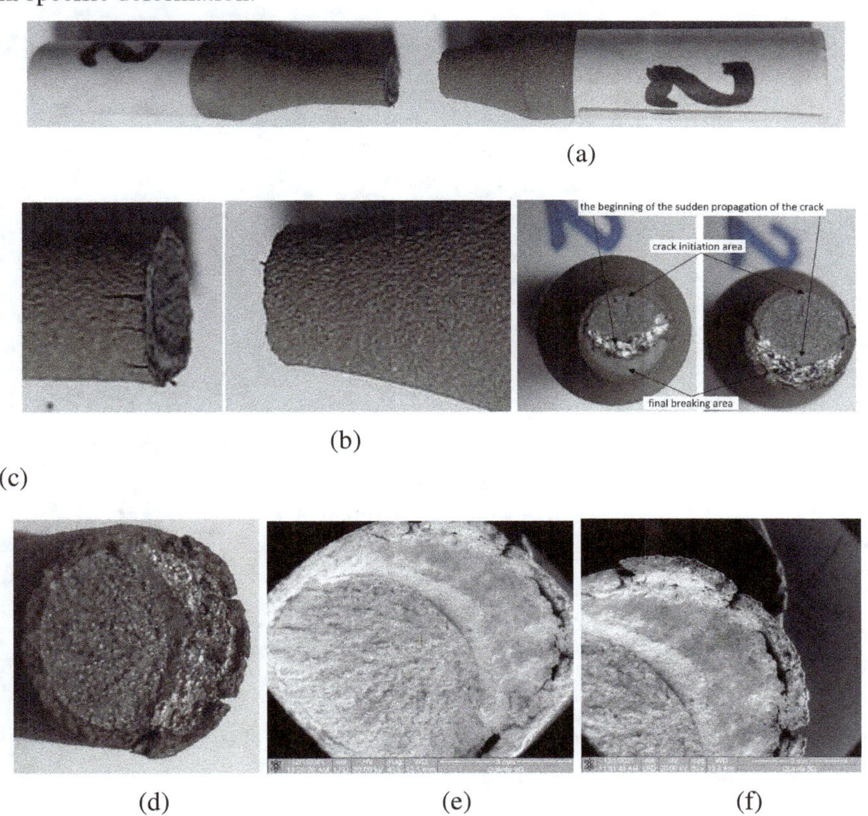

(a)

(b)

(c)

(d)   (e)   (f)

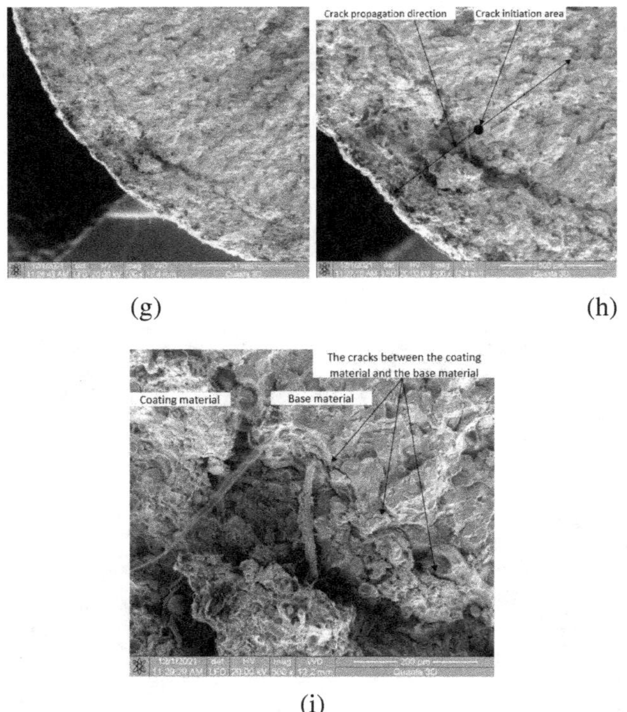

(g)   (h)

(i)

Figure 6.12 The macroscopic and microscopic appearance of the broken surfaces for sample 2,
σ = 636 MPa, N = 5300 cycles

Sample no. 3. Tension = 530 MPa. Number of cycles = 34660

Macroscopic observations

Sample 3 was subjected to a relatively high stress, approaching the yield strength (σ = 530 MPa). From Figure 6.13 (a), (b) and (c), it can be seen that at the end of the crack propagation period, the fracture surface changes its direction, from being perpendicular to the stress direction, to an angle of approx. 45^0. The explanations are as follows: during displacement in the base material, the crack front encounters a resistance to advance in the initial direction; given that a large part of the surface has been broken, the external force, which introduces high stresses, acts misaligned with respect to the surface left to break, Figure 6.13 (j). It should be noted that the stably propagated surface through fatigue, at low propagation speed, is smaller than the flat area, perpendicular to the direction of the forces, Figure 6.13 (c) and (d). In the region where the propagated crack front changes direction, significant detachments of the cover material from the base material are found, Figure 6.13 (d) and (e).

Microscopic observations

In the final fracture zone, large damages of the coating material are found, Figure 6.13 (f). For all samples, these large damages of the coating material in the final fracture zone are due to the large strain differences between the base material and the coating material. The base material having high deformation (in the plastic domain) in the final fracture zone also drives the cover material which, not being able to deform in the same way, is damaged by the appearance of large cracks. On the other hand, in the crack initiation area, Figure 6.13 (g), (h) and (i), no significant damage to the coating material is found. Of course, there is a detachment of the covering material from the base material which is due, here, too, to the differences in deformation in the elastic domain, Figure 6.13 (i).

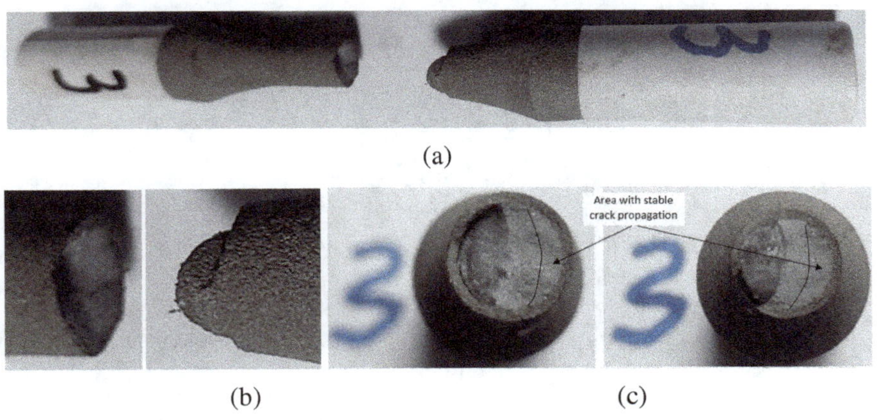

(a)

(b)   (c)

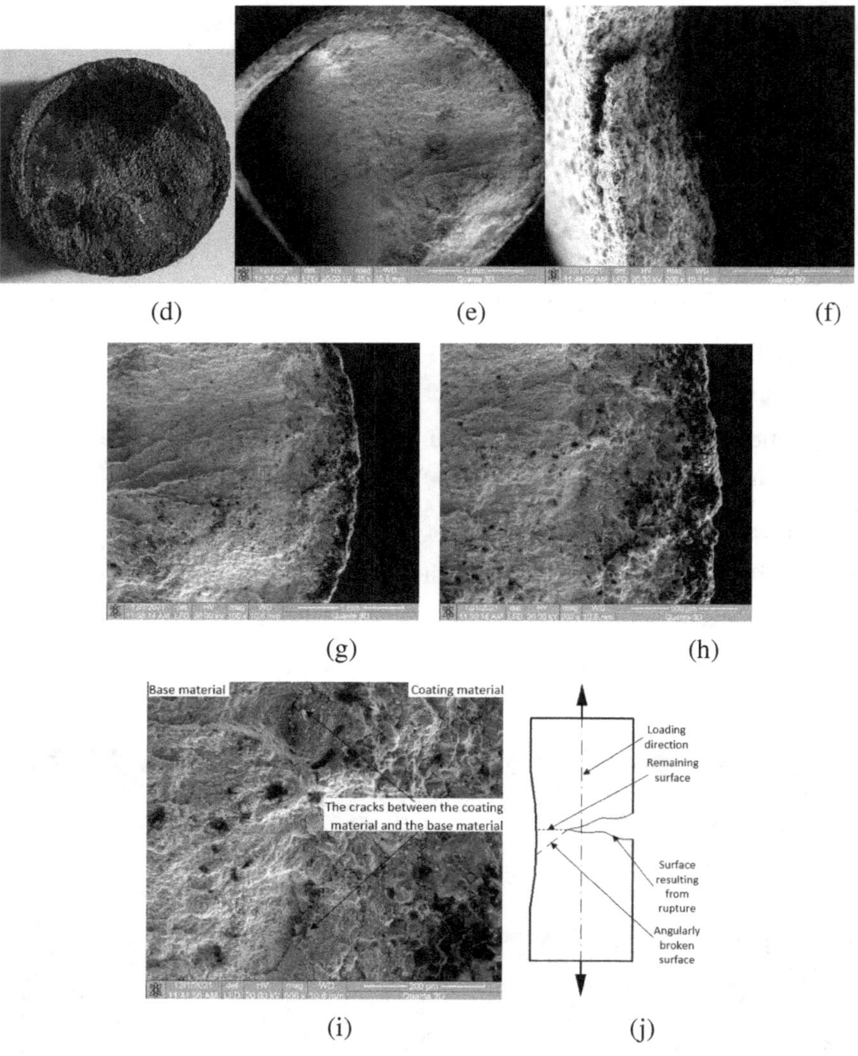

Figure 6.13 Macroscopic and microscopic appearance of broken surfaces for sample 3
σ = 530 MPa, N = 34660 cycles

Sample no. 6. Tension = 495 MPa. Number of cycles = 190000

Macroscopic observations

Sample 6 was subjected to a medium stress, relatively close to the yield point (σ = 495 MPa, σy=852 MPa). From Figure 6.14 (a), (b) and (c), it can be seen, as in the previous sample, that at the end of the crack propagation period, the fracture surface changes its direction, from being perpendicular to the stress direction at an angle of approx. 45^0. The explanations are the same as those provided for sample 3. Here, too, the stably propagated surface through fatigue, at low propagation speed, is smaller than the flat area, perpendicular to the direction of the forces, Figure 6.14 (c) and (d), (here the crack fronts were no longer drawn to distinguish more clearly the area between the two surfaces). The stress tension being lower, and, as a result, the deformations of the base material are also lower, no significant damage to the covering material is found in the area of the final break, Figure 6.14 (d), (e) and (f). Debonding of the cover material from the base material due to deflection of the crack surface naturally occurs.

Microscopic observations

In the crack initiation area, Figure 6.14 (g), (h) and (i), no significant damage to the coating material is observed. Of course, there is a detachment of the covering material from the base material which is due, here, too, to the differences in deformation in the elastic domain, Figure 6.14 (i).

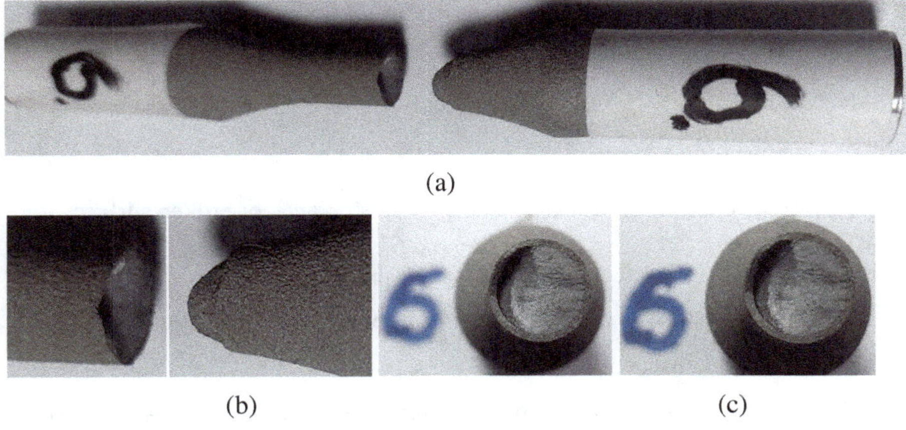

(a)

(b)   (c)

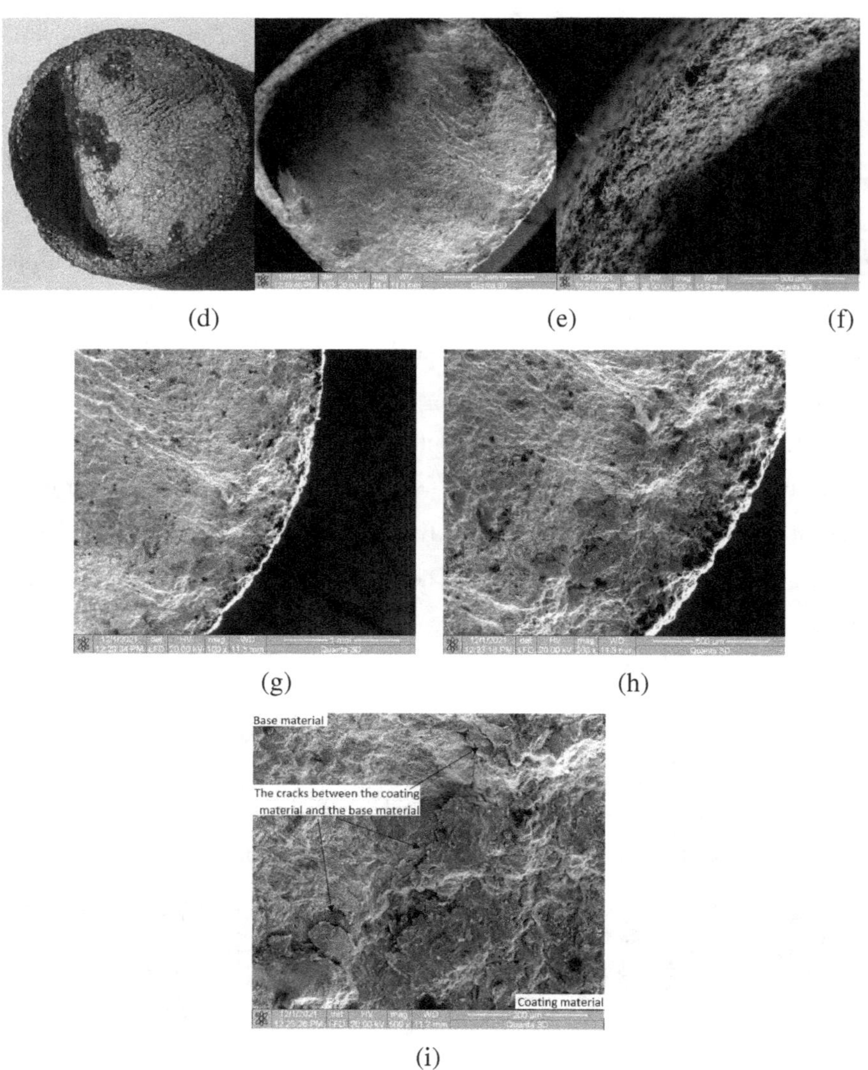

(d)　　　　　　　　(e)　　　　　　　　(f)

(g)　　　　　　　　(h)

(i)

Figure 6.14 Macroscopic and microscopic appearance of broken surfaces for sample 6

σ = 495 MPa, N = 190000 cycles

Sample no. 4. Tension = 477 MPa. Number of cycles = 878965

Macroscopic observations

The stresses to which the following samples are subjected, are getting lower. This also results in the fact that the fracture surfaces through stable fatigue crack propagation are increasingly larger and without deflection at the angle, Figure 6.15 (a), (b), (c) and (d). However, in this sample, the stress being higher than in the following samples and the area of stable crack propagation is smaller (see the corresponding figures from the following samples). Here the differentiation between the two zones of crack propagation is clearly observed: stable cyclic fatigue propagation – the area with the appearance of finer grains (crescent) and the zone of rapid (sudden) crack propagation, Figure 6.15 (c), (d) and (e). The application being at a lower voltage, no damage is observed by pulling out particles or by the appearance of large cracks, in the final rupture zone, Figure 6.15 (f).

Microscopic observations

In the fatigue crack initiation zone, no significant damage to the coating material is observed. Figure 6.15 (g). It is true, that at higher observation magnitudes, a detachment is observed that presents a radial crack through the coating material, Figure 6.15 (h) and (i). This is the very crack initiation zone, finding here a more weakly bonded portion of the coating material.

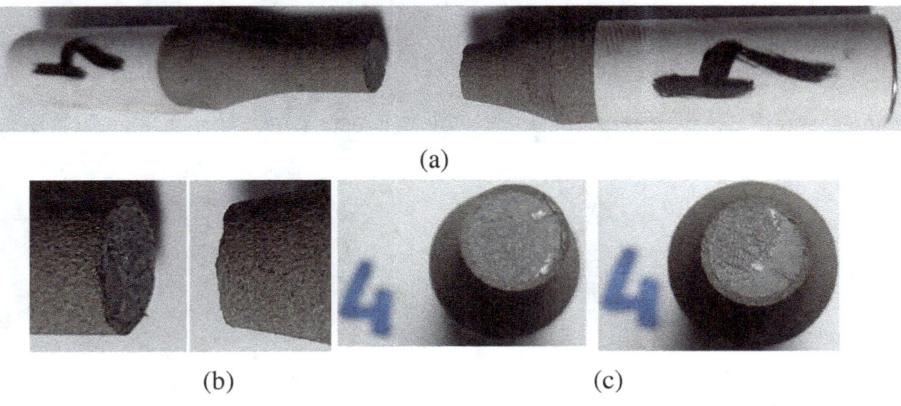

(a)

(b)   (c)

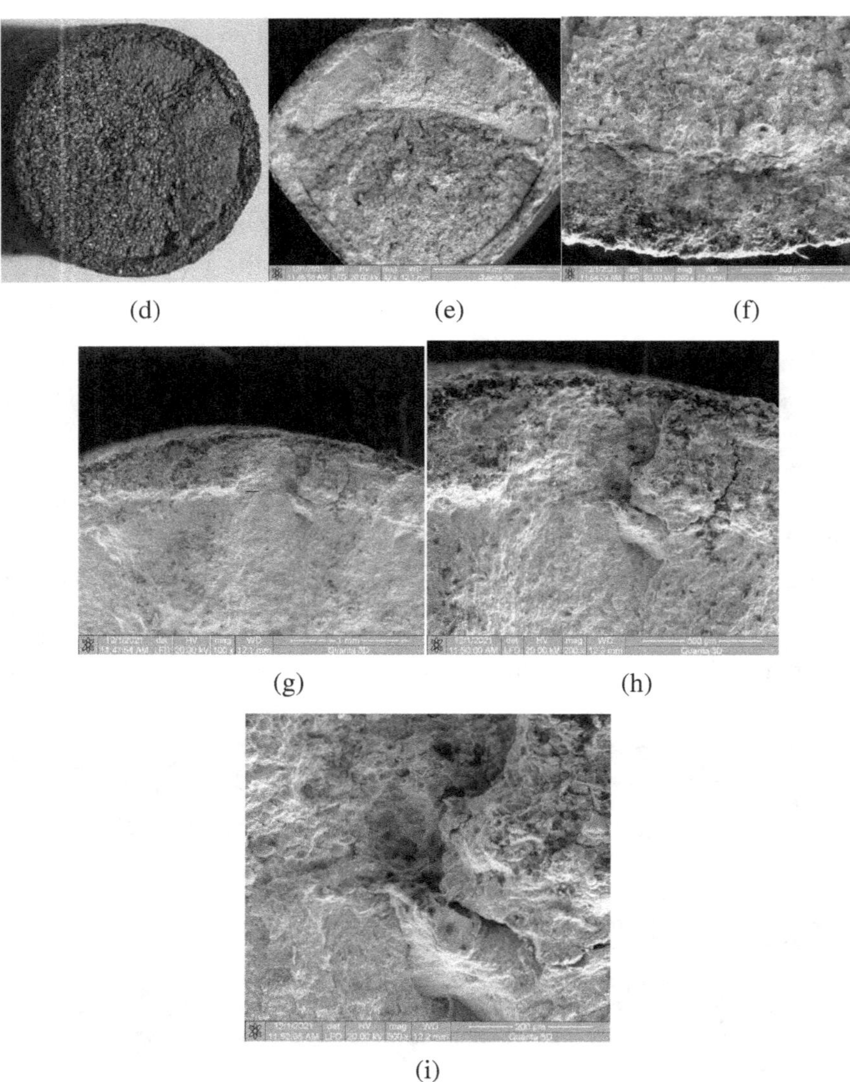

Figure 6.15 Macroscopic and microscopic appearance of broken surfaces for sample 4

$\sigma = 477$ MPa, N = 878965 cycles

Sample no. 7. Tension = 459 MPa. Number of cycles = 2287070

Macroscopic observations

The stress to which this specimen was subjected, is lower than that to which the previous specimen was subjected, hence the fracture surface by stable fatigue crack propagation is greater, Figures 6.16 (a), (b), (c) and (d). And here the differentiation between the two zones of crack propagation is clearly observed: the stable propagation through cyclic fatigue – the zone with the appearance of finer grains (in the form of a crescent) and the zone of rapid (sudden) propagation of the crack, Figure 6.16 (c), (d) and (e). The stress being at a lower voltage, it is found that the fracture surface belonging to the covering material is regular, perpendicular to the stress direction, following the crack front of the base material. As a result, at low voltages, as it will be seen in the following tests, it appears that the coating material does not acquire separate destructions from those of the base material, Figure 6.16 (d). In the zone of sudden crack propagation, Figure 6.16 (f), and in the zone of final rupture, no significant destruction of the covering material is observed.

Microscopic observations

In the fatigue crack initiation area, no significant damage to the coating material is found, Figure 6.16 (g). Of course, there is an area where a crack appears between the cover material and the base material, Figure 6.16 (h) and (i). This crack is caused by detachment occurring at the interface between the two materials as a result of differences in elastic deformation.

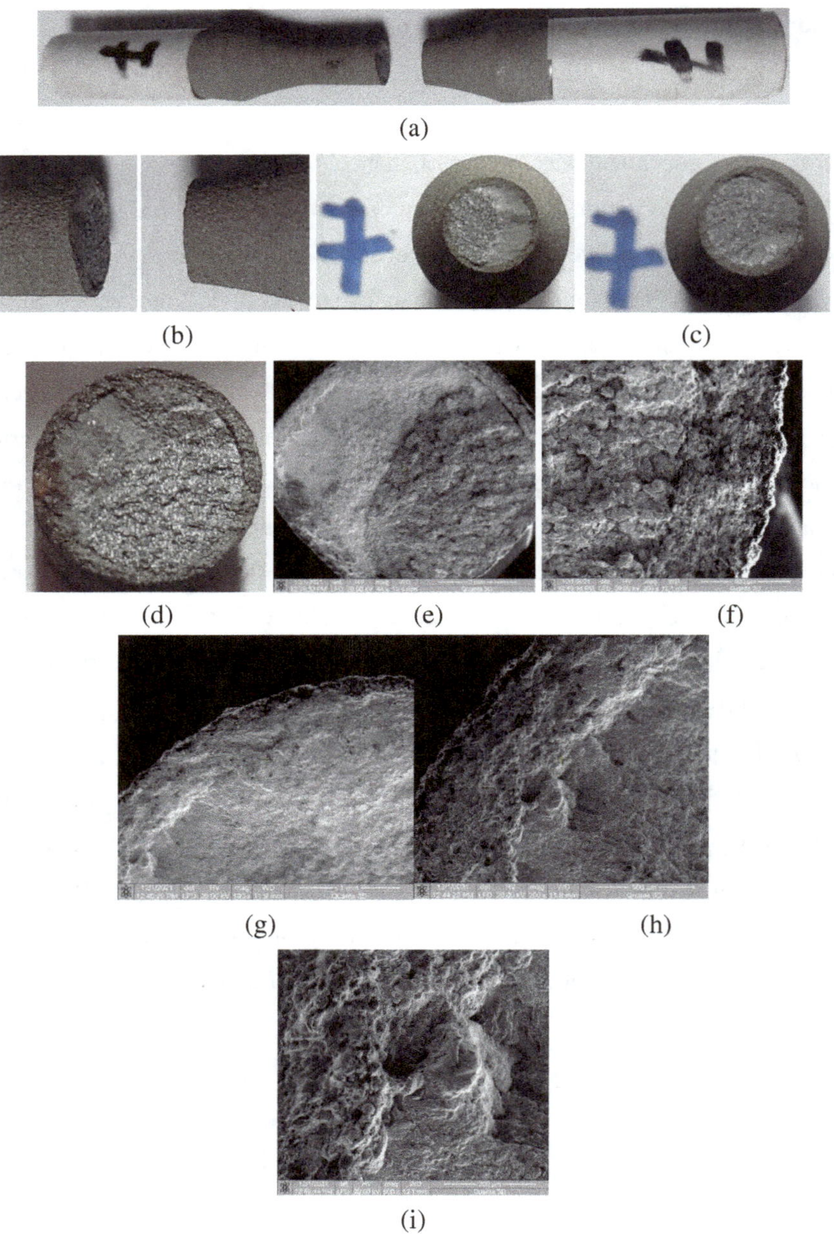

Figure 6.16 Macroscopic and microscopic appearance of broken surfaces for sample 7
$\sigma = 459$ MPa, N = 2287070 cycles

Sample no. 5. Tension = 449 MPa. Number of cycles = 3456873

Macroscopic observations

It is found that, in totality, the fracture surface is perpendicular to the stress direction, Figure 6.17 (a). As the stress to which this specimen was subjected is lower than that to which the previous specimen was subjected, the fracture surface obtained by stable fatigue crack propagation is larger, Figure 6.17 (b), (c) and (d). On the entire circumference of the tear surface belonging to the covering material, a macroscopic view shows no separation or significant damage of the covering material.

In the zone of sudden crack propagation, Figure 6.17 (e), and in the zone of final rupture, Figure 6.17 (e), no significant destruction of the covering material is found.

Microscopic observations

In the fatigue crack initiation zone, a very fine separation of the coating material from the base material is observed, Figure 6.17 (i). Small cracks are also observed in the area of the covering material. This happens as a result of the low crack propagation speed, in which case cracks can also occur in the covering material. At higher stresses, only the particles of the coating material detach, relative to each other, without time for inter- and intra-granular crack propagation in the coating material.

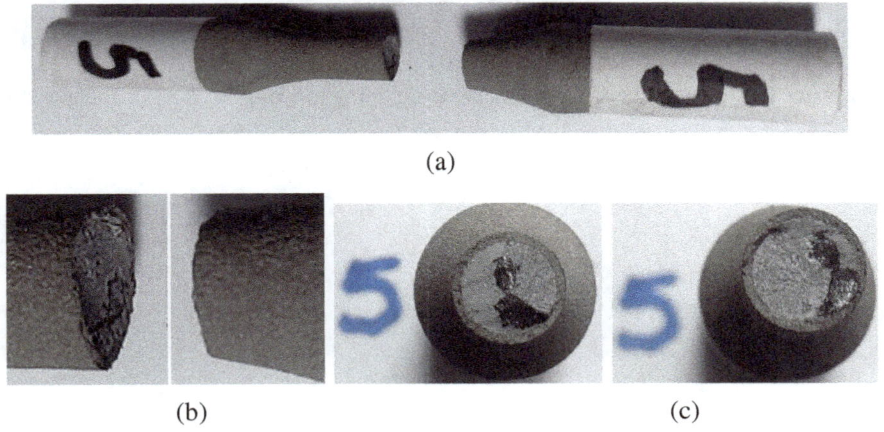

(a)

(b)   (c)

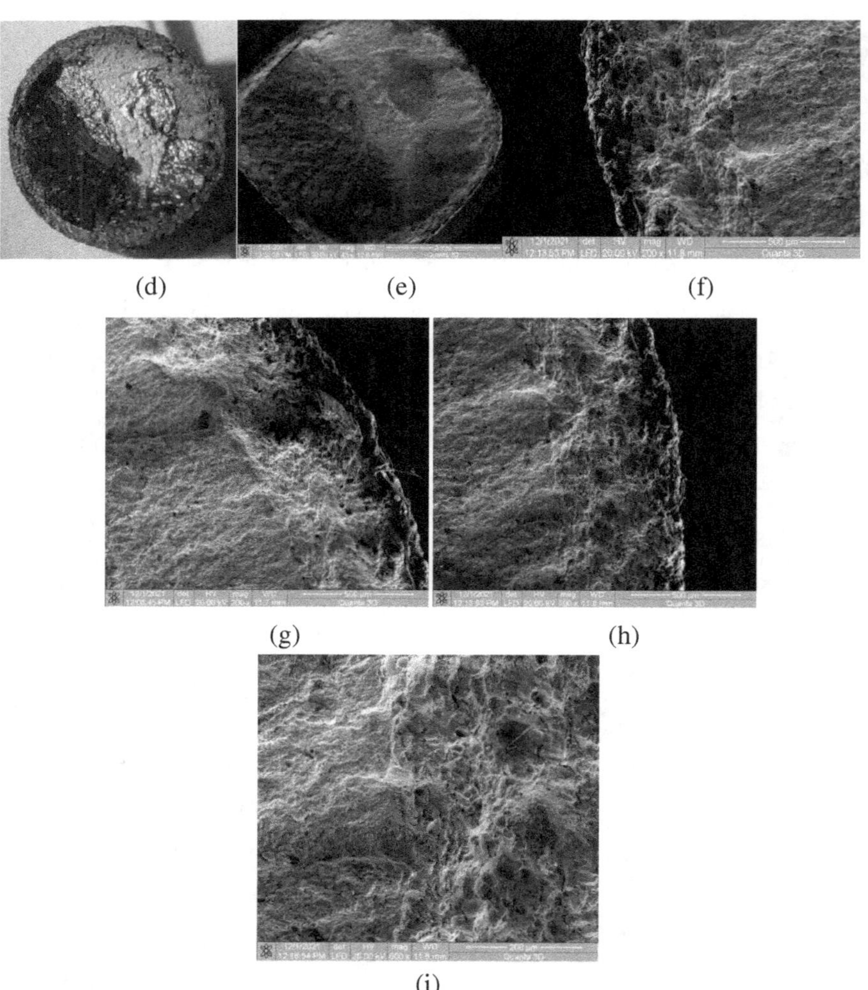

Figure 6.17 Macroscopic and microscopic appearance of broken surfaces for sample 5

σ = 449 MPa, N = 3456873 cycles

Sample no. 9. Tension = 446 MPa. Number of cycles = 4424120

Macroscopic observations

Considering the low stress to which it was subjected, more than 50% of the fracture surface is obtained by stable fatigue crack propagation. Here too, the fracture surface is found to be perpendicular to the stress direction, Figure 6.18 (a). It is also found that we have a uniformity in terms of the surface resulting from breaking the covering material, all around its circumference, Figure 6.18 (b), (c) and (e). A detachment of the cover material from the base material is found at the beginning of the sudden propagation of the crack front, Figure 6.18 (f).

Microscopic observations

In the fatigue crack initiation zone, no significant damage to the coating material is found, Figure 6.18 (g) and (h). As with the previous sample, a very fine crack is observed between the base material and the covering material, meaning a separation between the two materials. In this sample, inter- and intra-granular microcracks are observed, more pronounced than in the previous sample, Figure 6.18 (i), especially near the interface area between the two materials.

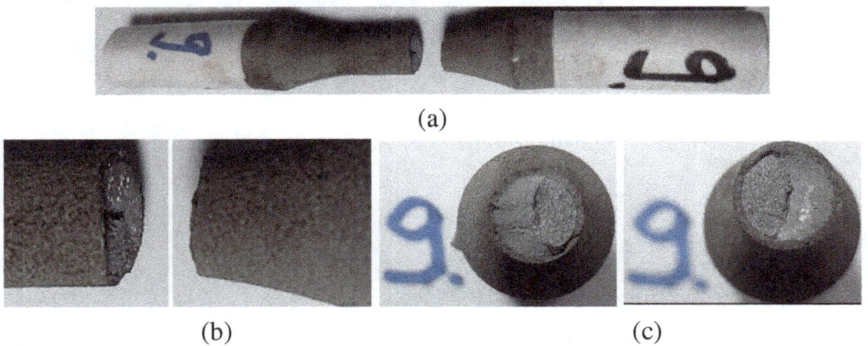

(a)

(b)   (c)

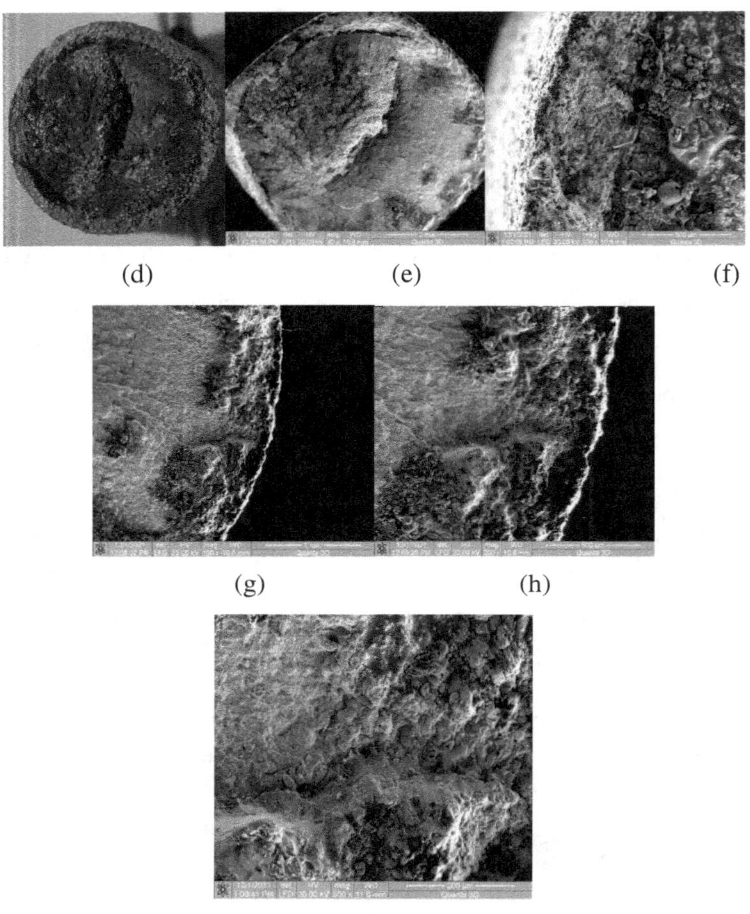

(d)   (e)   (f)

(g)   (h)

(i)

Figure 6.18 Macroscopic and microscopic appearance of broken surfaces for sample 9

$\sigma = 446$ MPa, N = 4424120 cycles

Sample no. 8. Tension = 438 MPa. Number of cycles = 5451948

As previously mentioned, the value of 5 million cycles was considered as the durability limit. Thus, if for a given stress, the specimen does not break after 5 million cycles, the test is stopped. This was the case of sample 8, which at the stress of 438 MPa, after 5451948 cycles did not break, and the testing was stopped. Carefully examining the outer surface of the sample, as a result, the covering material, no damage was found. The coefficient resulting from reporting the test stress to the yield point is: K=438/853=0.51. As a result, **at a demand voltage of approx. 50% of the value of the yield limit of the base material**, in conditions where cracking does not occur, the covering material resists high values of the number of cycles, without surface damage. This statement must be taken with caution, as we do not know what happens at the interface between the two materials, in the area where the first plastic deformations of the base material appear.

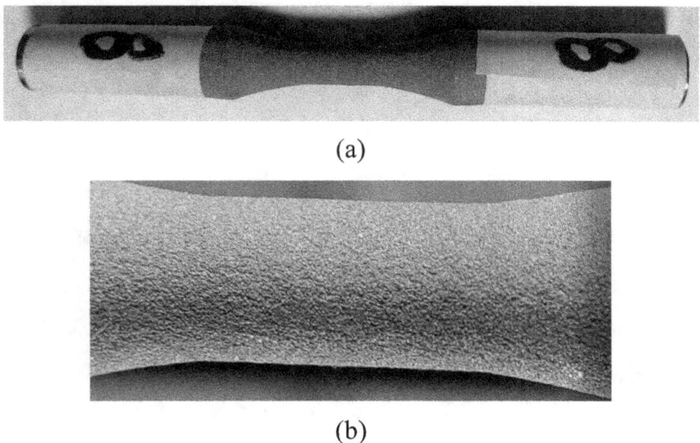

(a)

(b)

Figure 6.19 The macroscopic appearance of the outer surface of the sample 8
σ = 438 MPa, N = 5451948 cycles

After the tests, it was obtained with the help of data pairs, stress voltage, the number of cycles until breaking, the Wöhler diagram (Figure 6.20) for samples no. 8 (438MPa) and no. 1 (420MPa) that have not broken after more than 5 million cycles. It can be seen from the diagram that the fatigue limit is at a stress value of approx. 446 MPa.

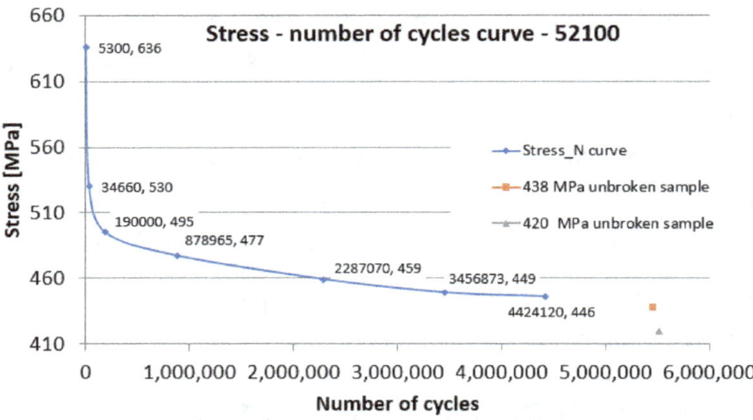

Figure 6.20 Wöhler diagram of the material with superficial deposits subjected to the fatigue test

6.2 Determination of hardness

Rockwell hardness tests were made up of the following specifications as follows:

- determination of Rockwell hardness with an applied force of 30 kgf (approx. 294N);

- testing of 1 uncoated sample (base material: AISI 52100 alloy steel);

- testing of 1 coated sample (deposits with Ni-Cr/C-based powders).

In Figure 6.21, both samples with the base material and samples with superficial deposits are shown after performing the hardness determination tests.

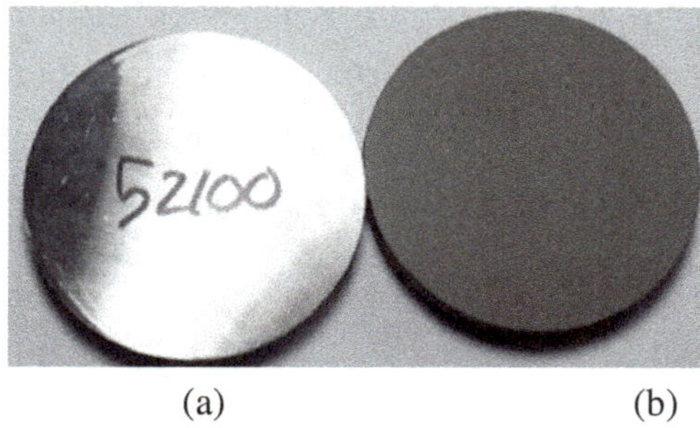

(a)   (b)

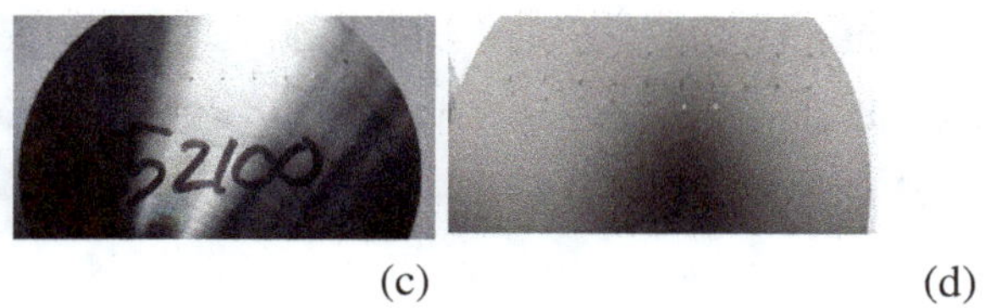

(c)   (d)

Figure 6.21 Images of the samples used in the hardness tests: (a, c) base material and (b, d) material with surface deposits

After carrying out the determinations, it was found that the deposited layer leads to an increase in hardness from approx. 28 HRC30 for the base material at approx. 45 HRC30 for the deposited layer.

Figure 6.22 shows the graph with the values obtained after determining the hardness, both for the base material and for the coating material.

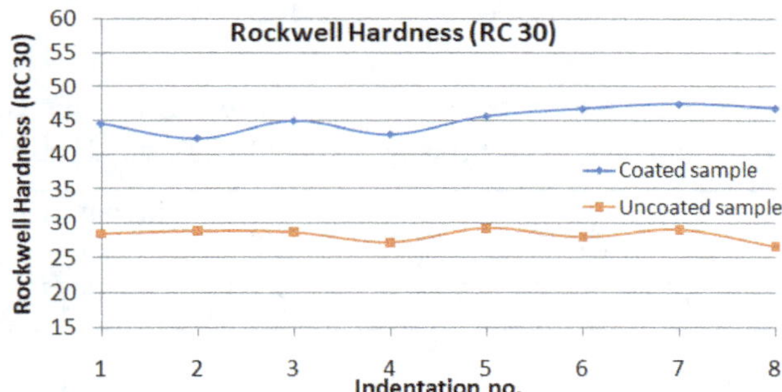

Figure 6.22 Rockwell hardness results for base material (blue line) and coated material (orange line)

6.3. Determination of the coefficient of friction by translational and rotational motion

To determine the coefficient of friction, tests were carried out by both translational and rotational motion.

Translational movement

The tests performed to determine the coefficient of friction through translational movement consisted of testing the following samples, as follows:

- 2 samples with superficial deposits subjected to an applied force of 10 N;

- 2 samples with superficial deposits subjected to an applied force of 20 N;

- 1 uncovered sample subjected to an applied force of 10 N;

- 1 uncoated sample subjected to an applied force of 20 N.

In testing, uncoated samples were tested for 15 minutes and coated samples for 60 minutes, at a linear speed of 10 mm/s.

The data obtained from the tests performed, are presented in Table 6.2, and the graphs resulting from the testing, can be seen in Figure 6.23, as follows: (a) sample 1-10 N for 60 minutes; (b) sample 2-10N for 60 minutes; (c) sample 3-20 N for 60 minutes; (d) sample 4-20 N for 60 minutes; (e) uncoated sample 1 10 N for 15 minutes; (f) uncoated sample 2 20 N for 15 minutes; (g) sample during testing.

Table 6.2 Data obtained from tests performed to determine the coefficient of friction by translational motion following the ASTM 132 standard

Sample	1	2	3	4	Uncoated 1	Uncoated 2
Load	10	10	20	20	10	20
COF	0.572	0.594	0.653	0.648	0.434	0.435

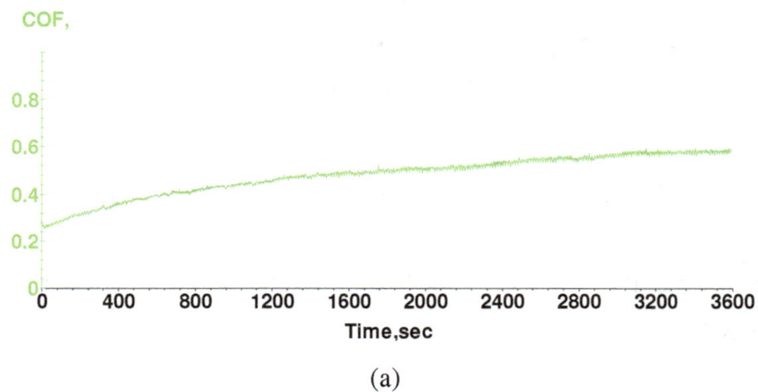

(a)

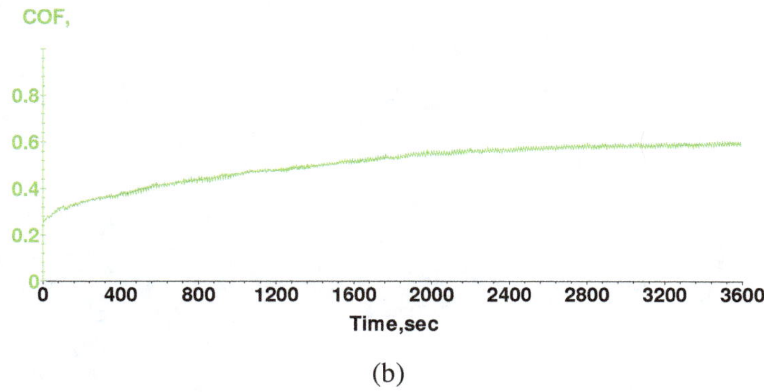

(b)

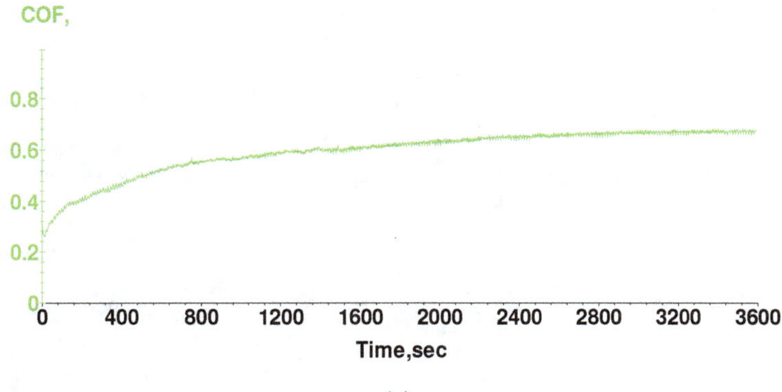

(c)

(d)

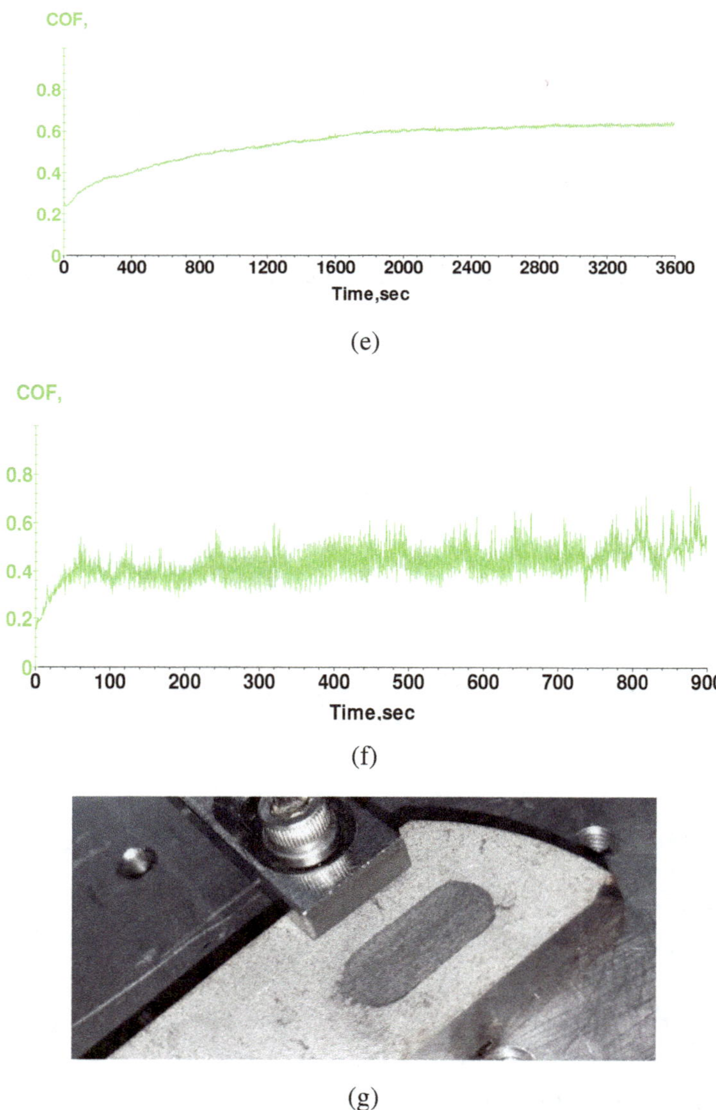

(e)

(f)

(g)

Figure 6.23 Graphs resulting from tests to determine the coefficient of friction by linear movement: (a) sample 1-10 N for 60 minutes; (b) sample 2-10N for 60 minutes; (c) sample 3-20 N for 60 minutes; (d) sample 4-20 N for 60 minutes; (e) uncoated sample 1 10 N for 15 minutes; (f) uncoated sample 2 20 N for 15 minutes; (g) sample during testing

In order to better highlight the data obtained for each individual sample, superimposed graphs were made according to Table 6.2 containing the values resulting from attempts to determine the coefficient of friction by translational movement. The superimposed graphs for the tests performed, are shown in Figure 6.24 as follows: (a) sample 1 and sample 2 at 10 N for 60 minutes; (b) sample 3 and sample 4 at 20 N for 60 minutes; (c) uncoated samples at 10 N and 20 N for 15 minutes.

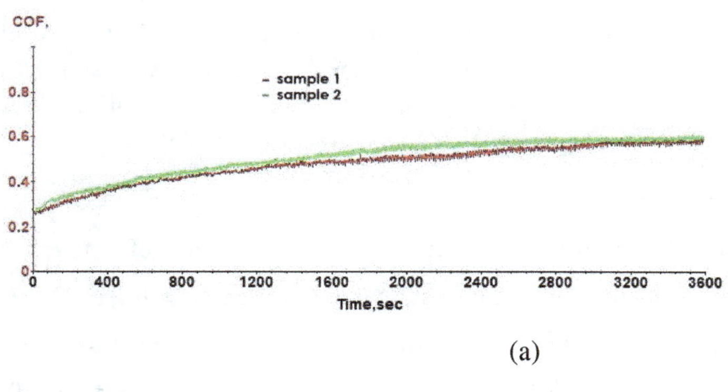

(a)

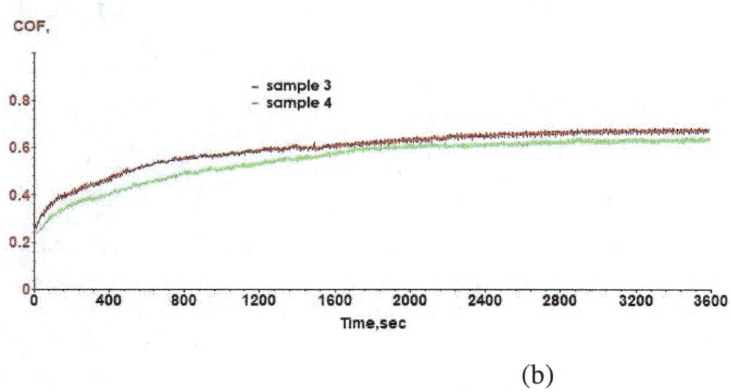

(b)

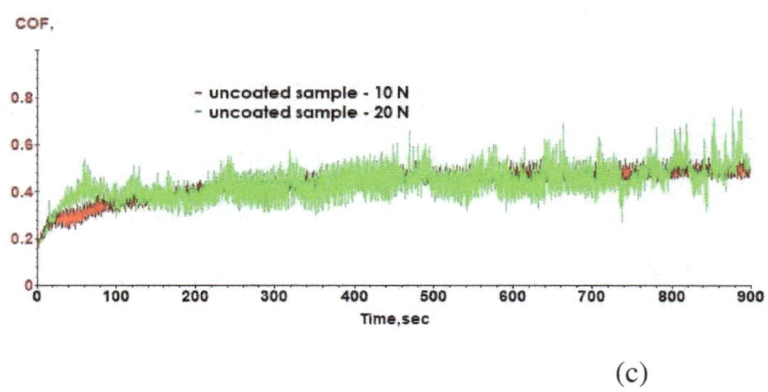

(c)

Figure 6.24 Superimposed plots for the tests performed on the determination of the coefficient of friction by linear motion: (a) sample 1 and sample 2 at 10 N for 60 minutes; (b) sample 3 and sample 4 at 20 N for 60 minutes; (c) uncoated samples at 10 N and 20 N for 15 min

Rotational movement:

The tests performed to determine the coefficient of friction through rotational movement consisted of testing the following samples, as follows:

- 2 samples with superficial deposits subjected to an applied force of 10 N;

- 2 samples with superficial deposits subjected to an applied force of 20 N;

- 1 uncoated sample subjected to an applied force of 10 N.

Uncoated samples were tested for 15 minutes and coated samples for 60 minutes, at a speed of 94.24 mm/s (linear speed).

The data obtained from the tests performed, are presented in Table 6.3, and the resulting graphs can be seen in Figure 6.25, as follows: (a) sample 1-10 N for 60 minutes; (b) sample 2-10 N for 60 minutes; (c) sample 3-20 N for 60 minutes; (d) sample 4-20 N for 60 minutes; (e) uncoated sample 10 N for 15 minutes; (f) sample during testing.

Table 6.3 Data obtained from tests performed to determine the coefficient of friction through rotational motion

Sample	1	2	3	4	Uncoated
Load	10	10	20	20	10
COF	0.614	0.643	0.668	0.687	0.196

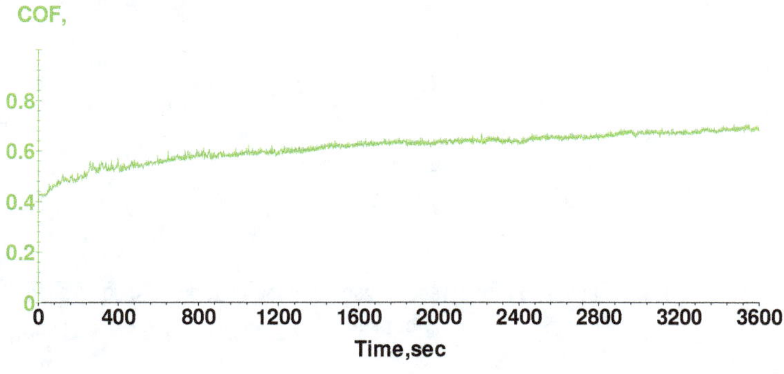

(a)

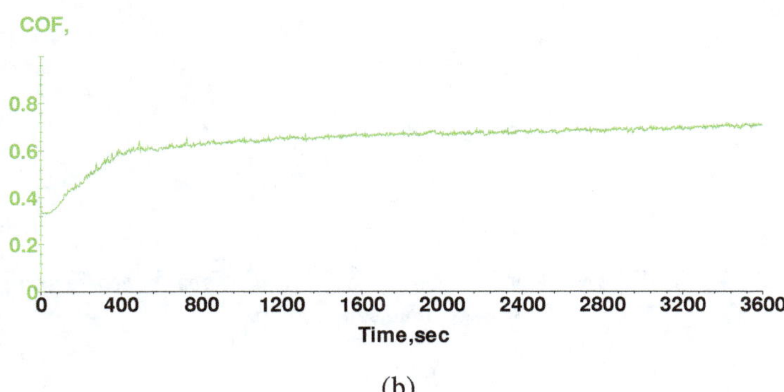

(b)

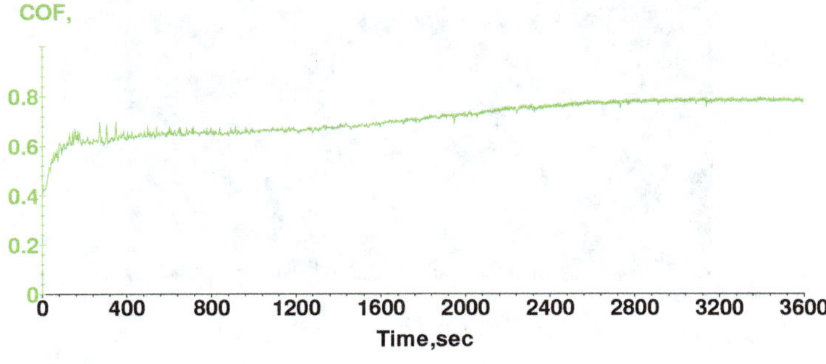

(c)

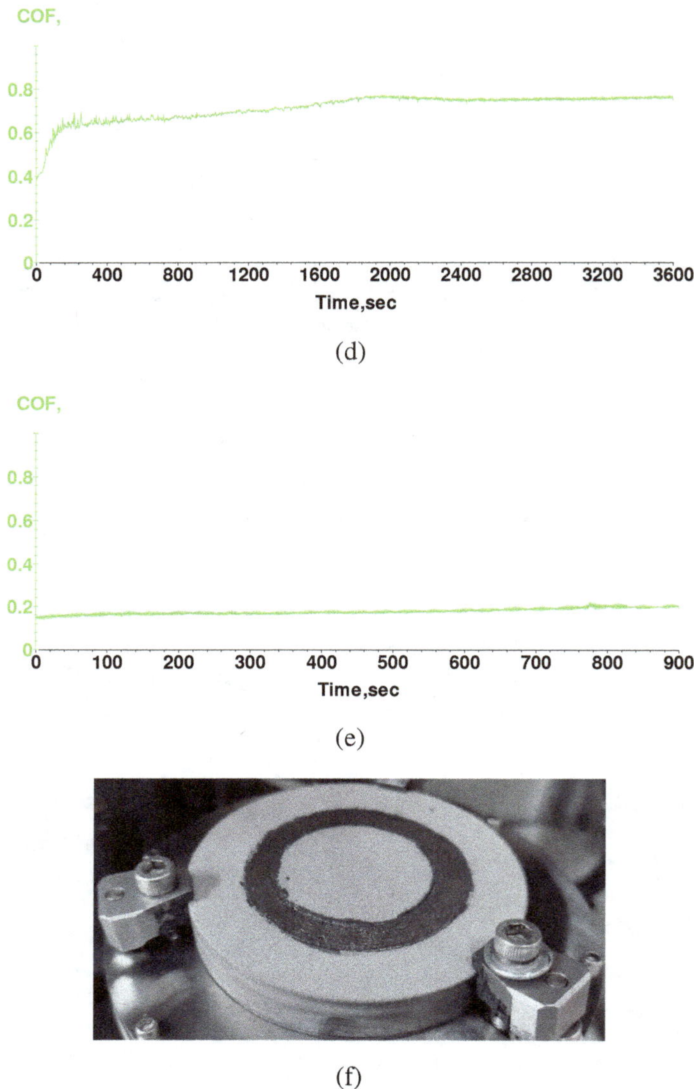

Figure 6.25 Resulting graphs for the determination of the coefficient of friction by rotational motion: (a) sample 1-10 N for 60 minutes; (b) sample 2-10 N for 60 minutes; (c) sample 3-20 N for 60 minutes; (d) sample 4-20 N for 60 minutes; (e) uncoated sample 10 N for 15 minutes; (f) sample during testing

In order to better highlight the data obtained for each individual sample, superimposed graphs were made according to Table 6.3, containing the values resulting from attempts to determine the coefficient of friction by rotational movement. The superimposed graphs for the tests performed, are shown in Figure 6.26 as follows: (a) sample 1 and sample 2 at 10 N for 60 minutes; (b) sample 3 and sample 4 at 20 N for 60 minutes; (c) uncoated sample at 10 N for 15 minutes.

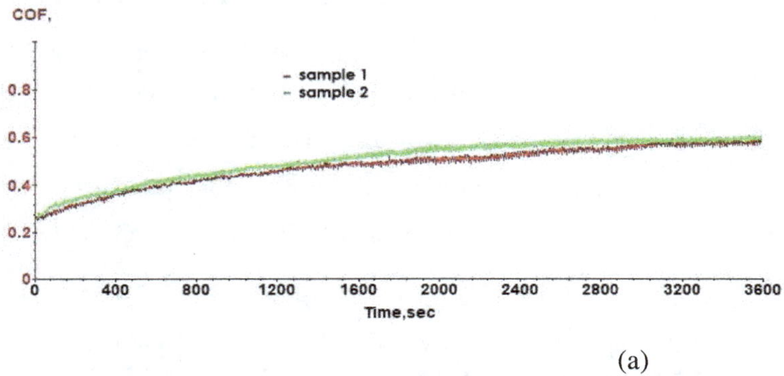

(a)

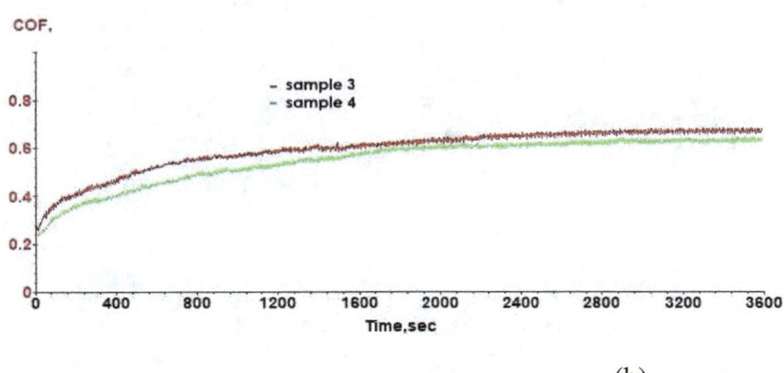

(b)

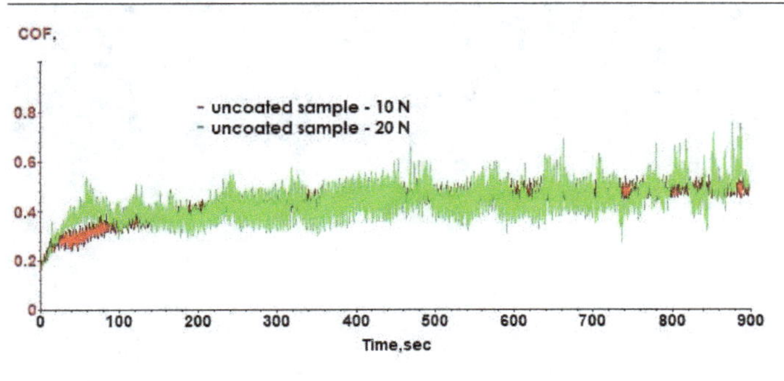

(c)

Figure 6.26 Superimposed plots for tests to determine the coefficient of friction by rotational motion: (a) sample 1 and sample 2 at 10 N for 60 minutes; (b) sample 3 and sample 4 at 20 N for 60 minutes; (c) uncoated sample at 10 N for 15 minutes

At the beginning of the coefficient of friction tests, a short transition period was observed, as the apparent contact area increased until it reached a steady state and the larger peaks of the coating surface asperities were removed. The last observation can explain the high values of the friction coefficient, which are between 0.4-0.7. Also, from the SEM images shown in Figure 6.27, it can be seen as well, the tips of the asperities are removed, and the particles left on the sliding track acted as an abrasive material, which leads to high values of the friction coefficient.

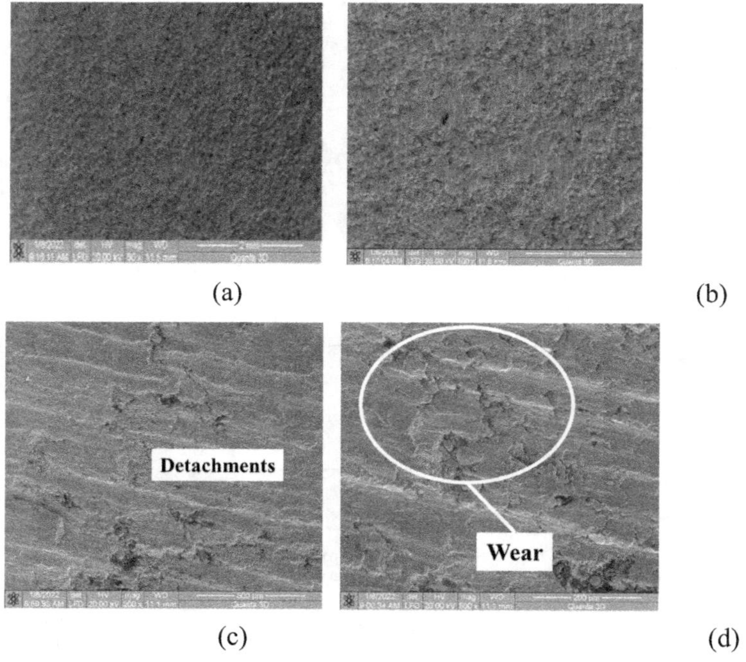

<table>
<tr><td>(a)</td><td>(b)</td></tr>
<tr><td>(c)</td><td>(d)</td></tr>
</table>

(e)

Figure 6.27 Images at different magnifications of the sample surface after friction coefficient determination tests (a) 50X; (b) 100X; (c) 200X; (d) 500X; (e) 1000X

6.4. Microindentation tests and determination of layer adhesion by microscratch

Microindentation:

To determine the hardness, 5 samples were tested, of which 4 samples with superficial deposits and one uncoated sample. The microindentation tests constituted for all the samples, the submission to a force of 20 N, they were performed 5 times, and the final data are presented in Table 6.4.

Table 6.4 Data obtained after performing microindentation tests

Sample		1	2	3	4	Uncoated
Load [N]				20		20
	1	69.60	65.80	72.23	70.44	49.31
	2	68.36	61.74	72.97	59.37	50.96
Rockwell Hardness	3	73.87	68.07	67.19	65.95	47.28
[HRC]	4	70.72	69.23	70.97	71.64	52.45
	5	63.88	64.79	76.26	69.03	48.59
	Av	**69.29**	**65.92**	**71.92**	**67.29**	**49.72**
	1	155.47	152.82	162.22	166.88	199.86
	2	154.38	159.67	169.53	153.40	208.04
Young Modulus	3	160.55	160.00	165.44	162.03	221.24
[GPa]	4	155.92	157.04	160.25	165.28	211.39
	5	166.72	160.88	168.86	163.77	209.69
	Av	**158.61**	**158.08**	**165.26**	**162.27**	**210.04**

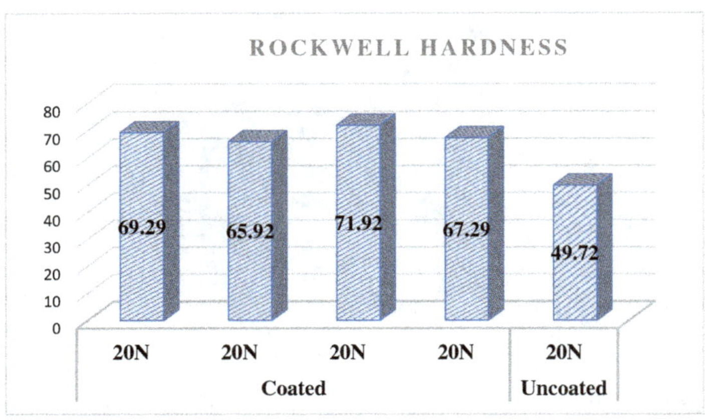

Figure 6.28 Comparative graph of the results obtained from the microindentation tests

The graphs resulting from the microindentation tests show the specific load and penetration depth, as follows in Figure 6.29: sample 1-20N (a); sample 2-20N (b); sample 3-20N (c); sample 4-20N (d); uncovered-20N (e).

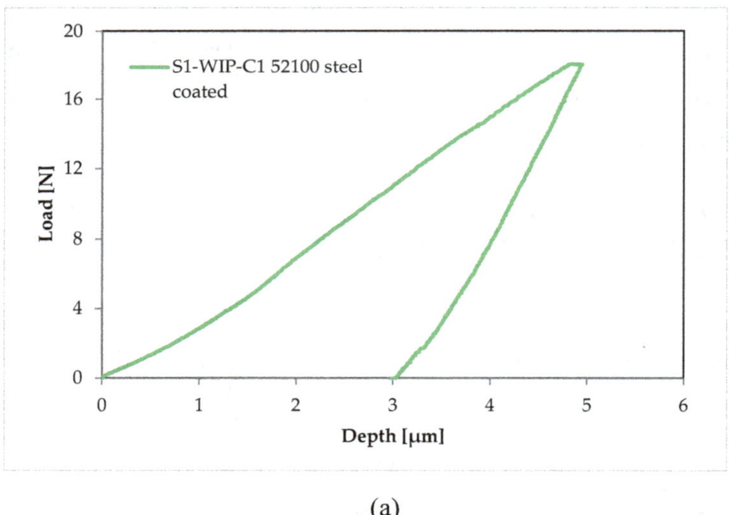

(a)

20
16
12
8
4
0
Load [N]
S2-WIP-C1 52100 steel
coated
0
1
2
3
4
5
6
Depth [μm]

(b)

20
16
12
8
4
0
Load [N]
S3-WIP-C1 52100 steel
coated
0
1
2
3
4
5
6
Depth [μm]

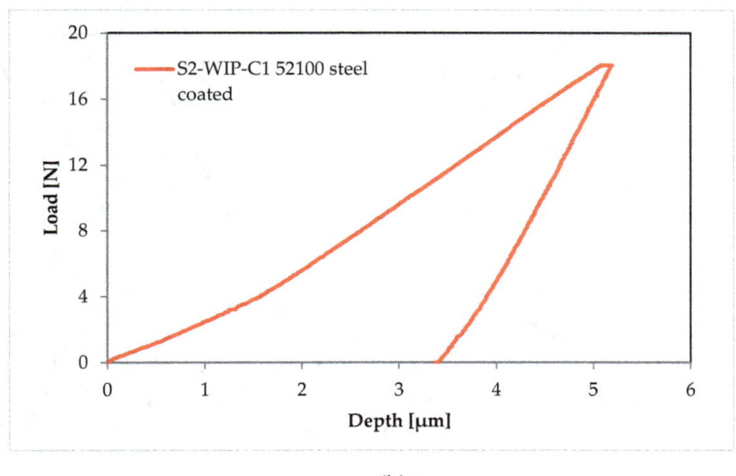

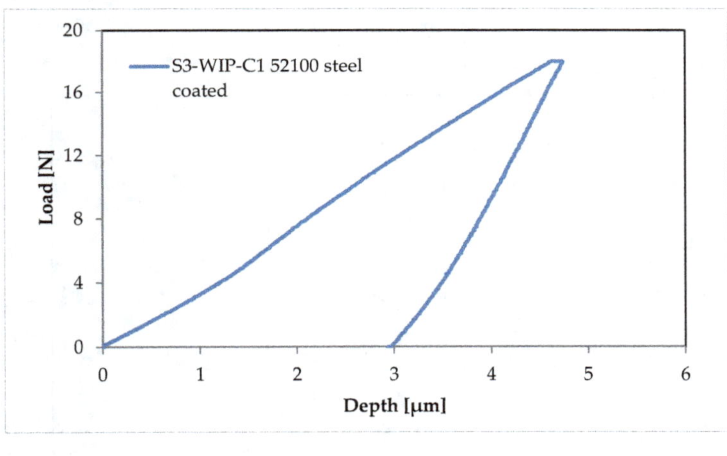

(c)

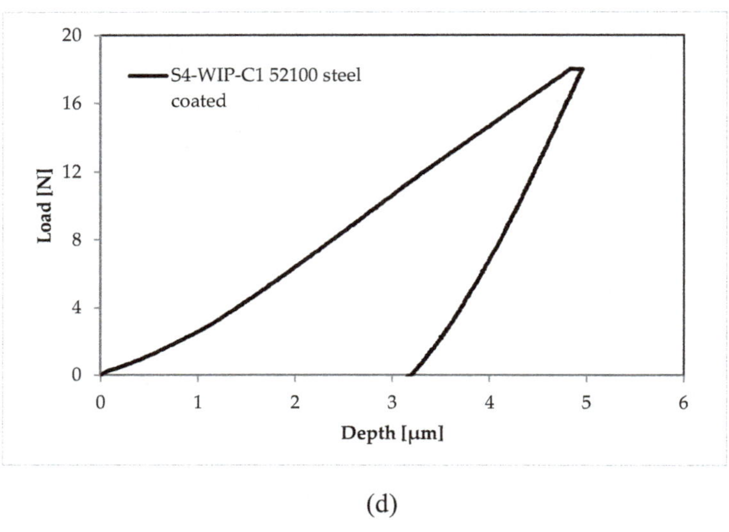

(d)

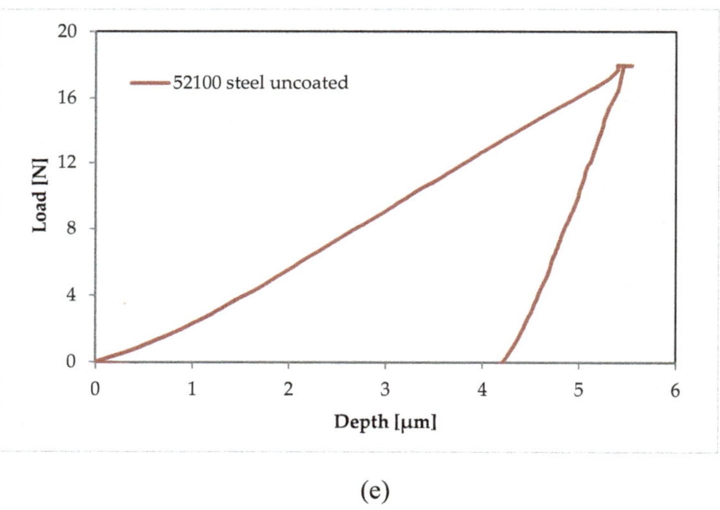

(e)

Figure 6.29 Graph of applied force and penetration depth: sample 1-20N (a); sample 2-20N (b); sample 3-20N (c); sample 4-20N (d); uncovered-20N(e)

Figure 6.30 shows the comparative graphs following microindentation tests for the determination of Rockwell hardness with applied force (20 N) and penetration depth for both coated and uncoated samples. Coated samples are represented by green, red, blue and black, and uncoated sample is represented by cherry.

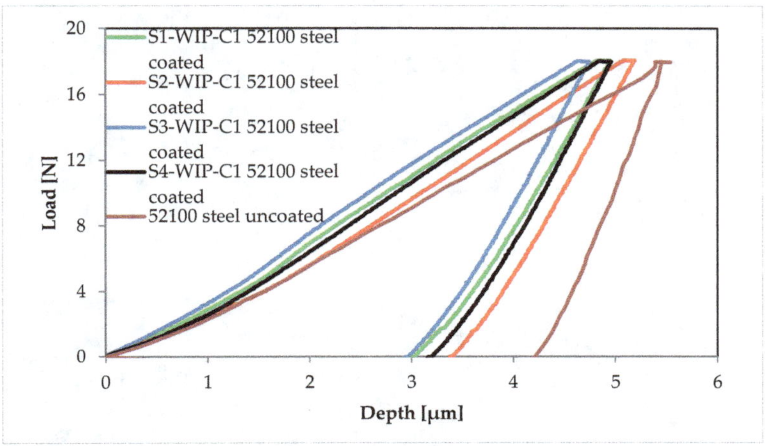

Figure 6.30 Comparative plots of applied force and penetration depth for both coated and uncoated samples at 20 N applied force

Figure 6.31 shows typical microindentation marks at 10 N and 20 N, obtained by SEM. The plastic deformation of the deposited layer is observed, but no cracks or material damage were identified in the structure.

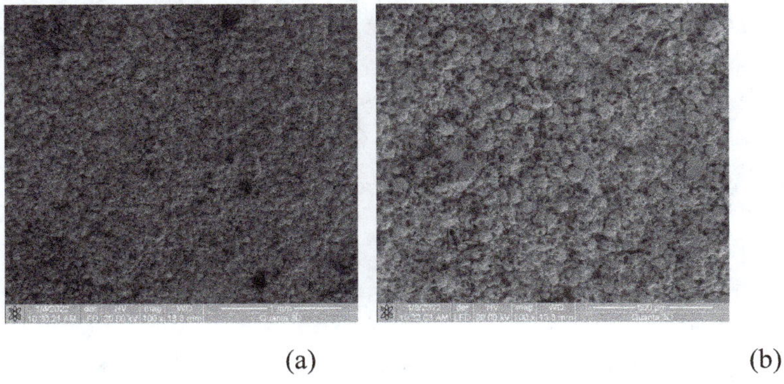

(a)   (b)

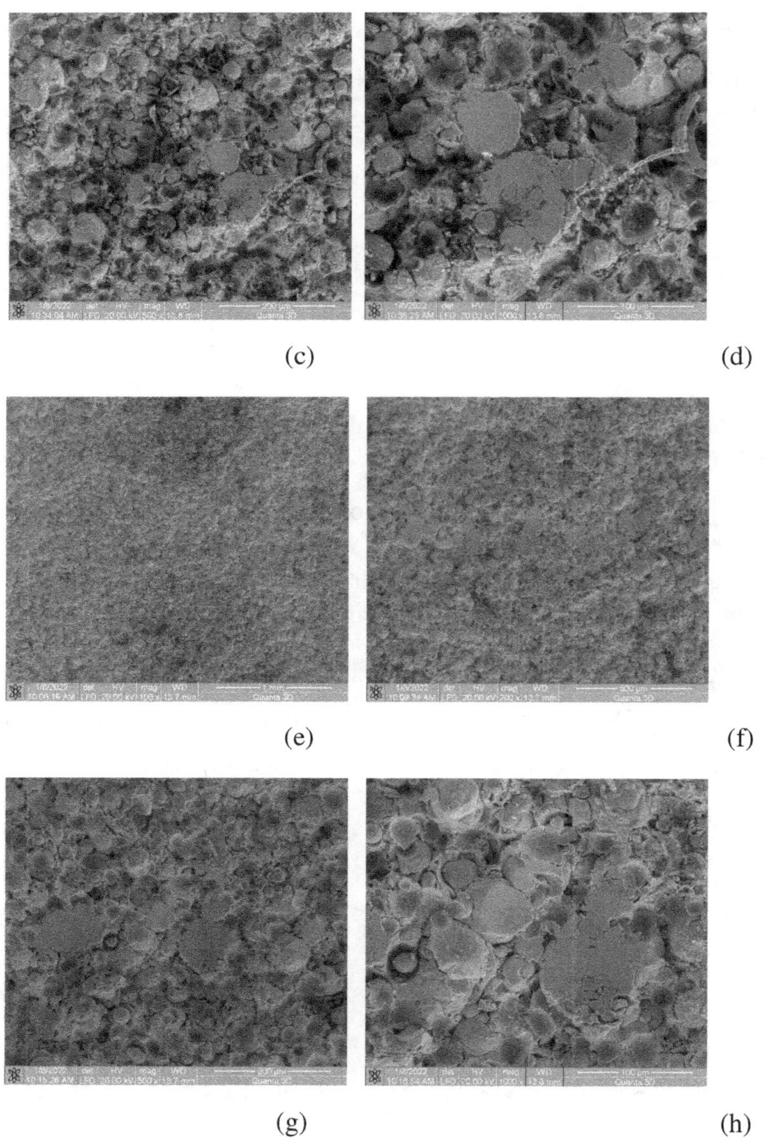

(c)   (d)

(e)   (f)

(g)   (h)

Figure 6.31 SEM images of indentation marks at a load of 10 N (a) 100X; (b) 200X; (c) 500X; (d) 1000X and 20 N (e) 100X; (f) 200X; (g) 500X; (h) 1000X

Microscratch:

The tests performed to determine the resistance to microscratches consisted of testing the following samples, as follows:

- 5 samples at a linear load from 1 to 10 N;

- 5 samples at a constant load of 10 N;

- the test duration is 60 s.

The values generated as a result of testing the resistance of the samples to microscratches are presented in the following table (Table 6.5.):

Table 6.5. Data obtained from micro-scratch resistance testing following ASTM D7187

Sample	1	2	3	4
Linear load [N]				
COF	0.255	0.213	0.484	0.379
Constant load[N]				
COF	0.329	0.347	0.559	0.615

By means of the following figures, the graphs obtained from the tests carried out, are exposed where it can be observed the variation of the coefficient of apparent friction COF with the displacement Y which presents many peaks and have average values greater than 0.2, and at first glance, all the samples have a resistance good for microscratching. Correlation with SEM analyses presents a better exposition of the obtained results. Figure 6.32 shows the graphs resulting from the tests for linear load and constant load, as follows: (a) sample 1-10N linear load; (b) sample 2-10N linear load; (c) sample 3-10N linear load; (d) 4-10N linear load test; (e) test 1-10N constant load; (f) test 2-10N constant load; (g) test 3-10N constant load; (h) test 4-10N constant load.

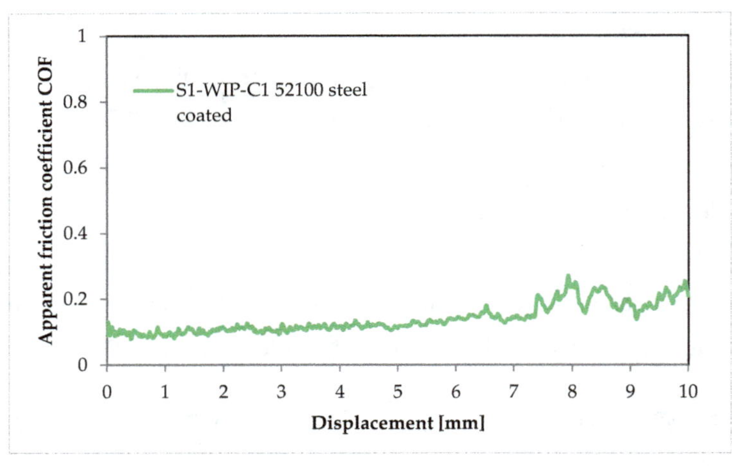

(a)

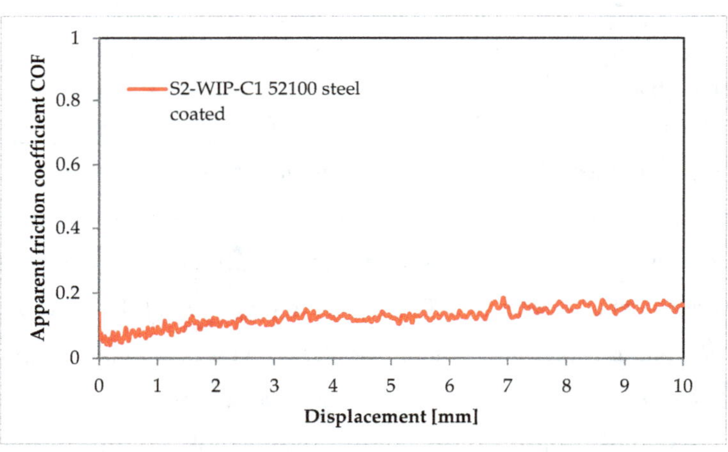

(b)

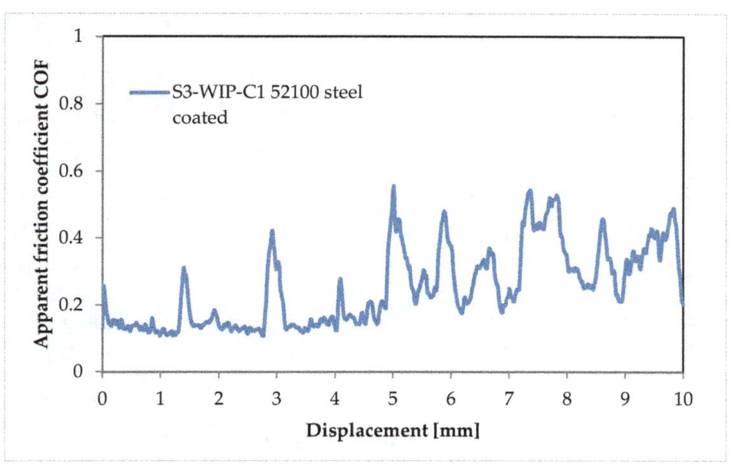

(c)

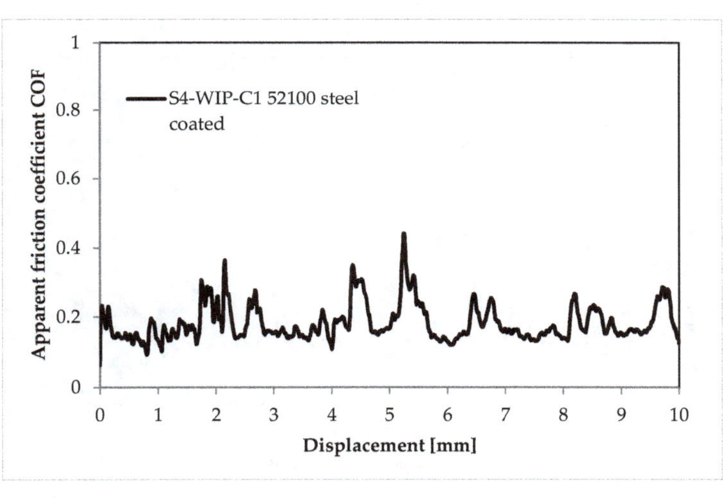

(d)

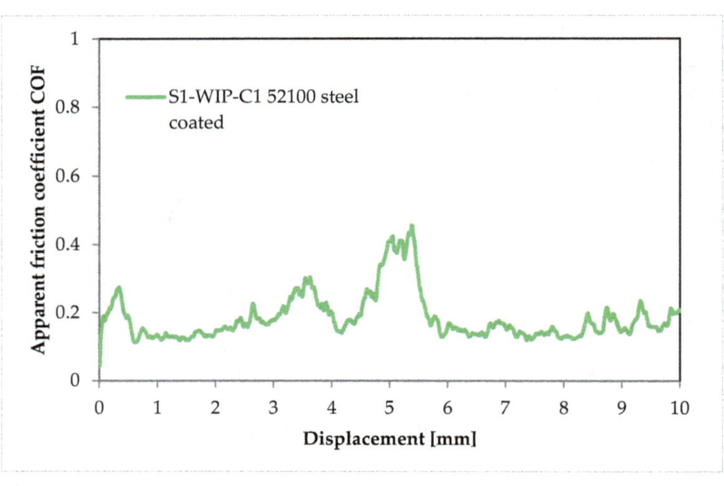

(e)

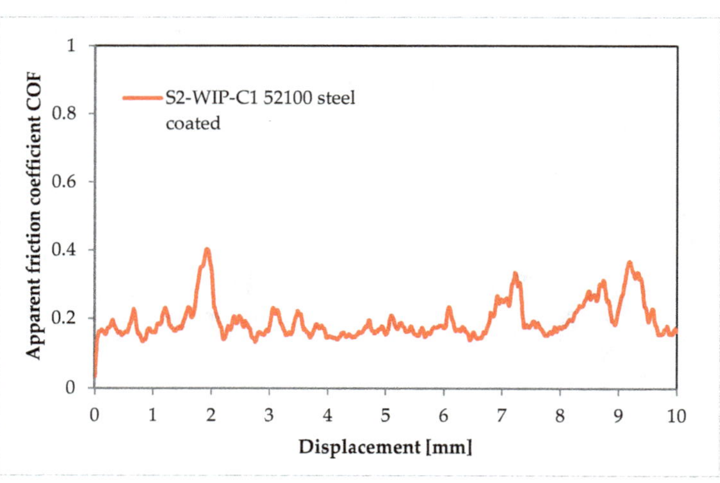

(f)

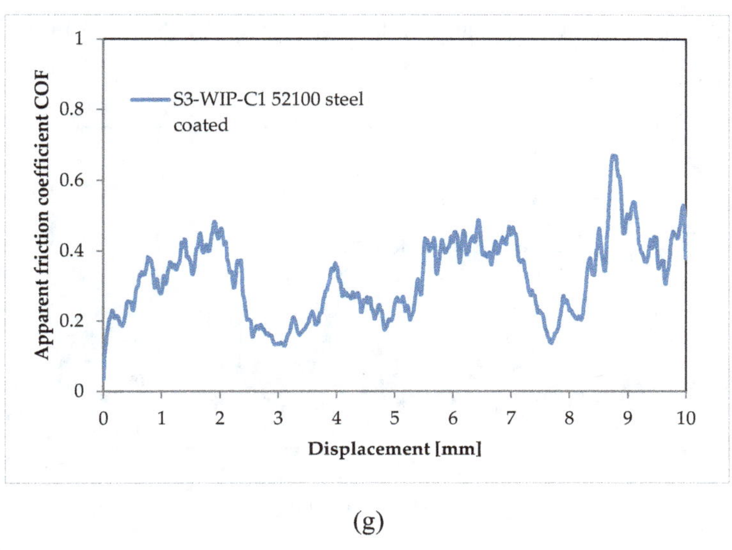

(g)

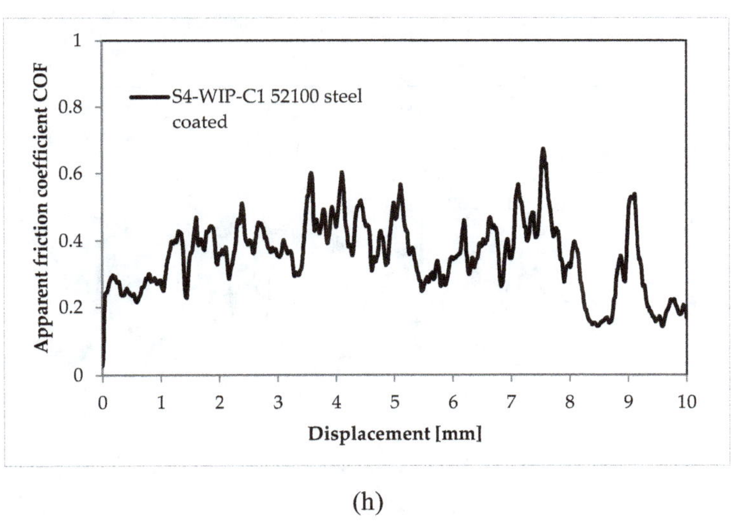

(h)

Figure 6.32 Graphs obtained after determining the resistance to microscratches: (a) sample 1-10N linear load; (b) sample 2-10N linear load; (c) sample 3-10N linear load; (d) 4-10N linear load test; (e) test 1-10N constant load; (f) test 2-10N constant load; (g) test 3-10N constant load; (h) test 4-10N constant load

Figure 6.33 shows comparative graphs following the tests performed to determine the resistance to microscratches at linear load and constant load (sample 1 green colour; sample 2 red colour; sample 3 blue colour; sample 4 black colour), as follows: (a) samples for linear load at 10N and (b) samples for constant load at 10N.

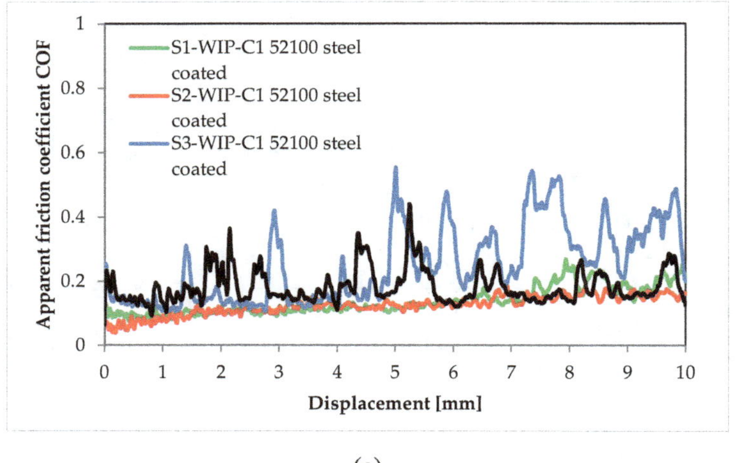

(a)

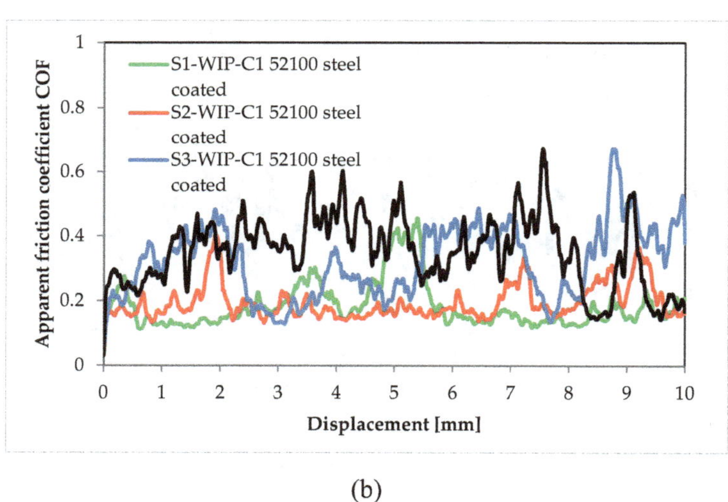

(b)

Figure 6.33 Comparative graphs resulting from the determination of microscratch resistance: (a) samples for linear load from 1 to 10N and (b) samples for constant load at 10N

At first glance, the samples show good microscratch resistance, because the variation of apparent friction coefficient COF, with Y displacement shows many peaks. From the SEM images shown in Figure 6.34, at 200X, it can be seen that for linear loading, the width of the microscratch track increases with increasing loading and for constant loading, the width of the microscratch track remains the same at about 260 µm. On the other hand, from the 500X and 1000X magnification SEM images, it can be seen that the microscratch track is only on the coating material and the microscratch blade did not touch the base material. We can see that the microscratch blade only removes the large asperities of the surface of the coating material and this shows us a very good microscratch resistance of the coatings.

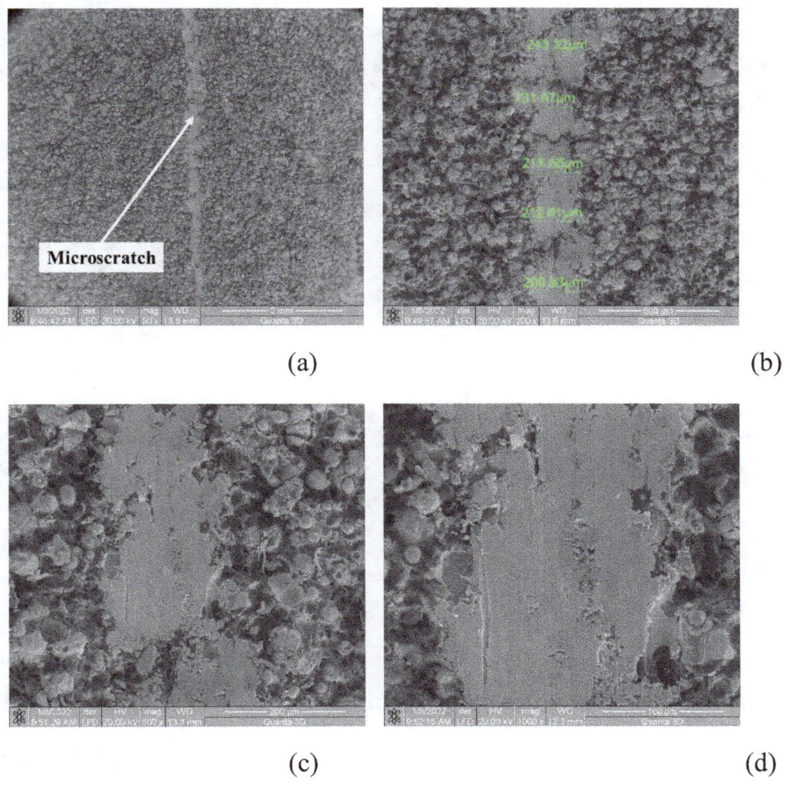

(a)

(b)

(c)

(d)

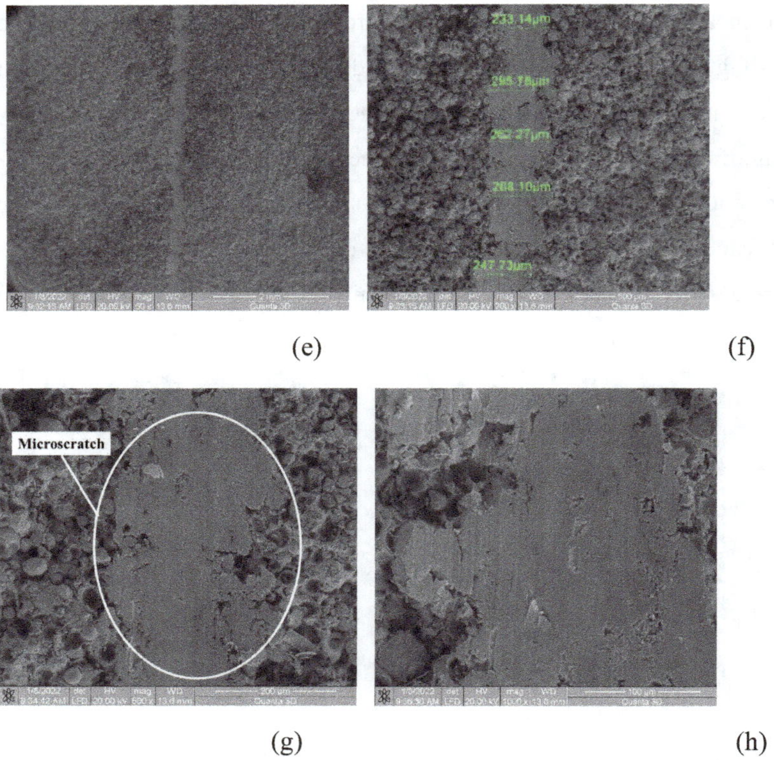

(e) (f)

(g) (h)

Figure 6.34 SEM images following microscratch testing at a linear load (a) 50X; (b) 200X; (c) 500X; (d) 1000X; and constant (e) 50X ; (f) 200X; (g) 500X; (h) 1000X

6.5 Partial conclusions

1. Following the fatigue tests, the macroscopic and microscopic analysis could highlight the area of long-term fatigue propagation and the area of sudden crack propagation followed by sample rupture; the crack propagation zone is characterised by a fine structure specific to the sudden damage determined by the coarser particles; it was found that there are no significant pores or cracks at the level between the covering material and the base material; in the crack initiation phase, no detachments of the covering material due to it were observed; test no. 8, managed to exceed the reference value so that the test was

stopped at a stress value of 438 MPa and a no. of 5451948 cycles, without observing damage to the deposited material or its detachment, but it should be noted that the condition of the surfaces between the base material and the coating where the first deformations appear at the level of the base material can not be appreciated.

2. The hardness of the material increased from 28 HRC for the base material to 45 HRC for the deposited layer.

3. Microscratch tests showed that the Ni-CrC layer has good microscratch resistance, as the friction coefficient showed a large variability along the sliding path with average values greater than 0.2, and the blade only removes the large asperities of the coating, which confirms the achievement of good resistance to microscratches.

4. The apparent coefficient of friction of the coatings was determined by both linear and rotational motion. A short transition period was observed at the beginning of both tests, as the apparent contact area increased until it reached a steady state and flattened the large peaks of the coating asperities. Free particles increase the coefficient of friction, explaining the COF values between 0.4 and 0.7.

CHAPTER 7 Experimental results on the corrosion behaviour of deposits by the Cold Spray method

7.1 Determination of electrochemical corrosion behaviour by the potentiodynamic method

Research on electrochemical corrosion processes by potentiodynamic testing enables instantaneous information on corrosion rates, linear bias resistance and Tafel extrapolation techniques.

A significant improvement in corrosion resistance can be seen. Figure 7.1 shows the potentiodynamic polarization plots for coated and uncoated 52100 sample for comparison of results, in HNO_3-H_2SO_4 acid rain solution. As it can be seen from the graphs, the coated 52100 sample shows lower icorr and higher E0 than the uncoated 52100, due to the effect of the electrolyte being blocked by surface deposits. As it can be seen from Table 7.1, the icorr value of 18.37 mA for the substrate is the highest. This difference is also confirmed by the corrosion rate values of 88.48mpy for the uncoated 52100 compared to 18.95mpy for the coated 52100 sample. This lower rate is due to the surface condition of the sample, a flat surface without cracks, confirmed in the microstructural analysis. Due to the nature of the coating materials, chromium and Ni carbide, these coatings exhibit excellent chemical stability and corrosion resistance in most aggressive environments.

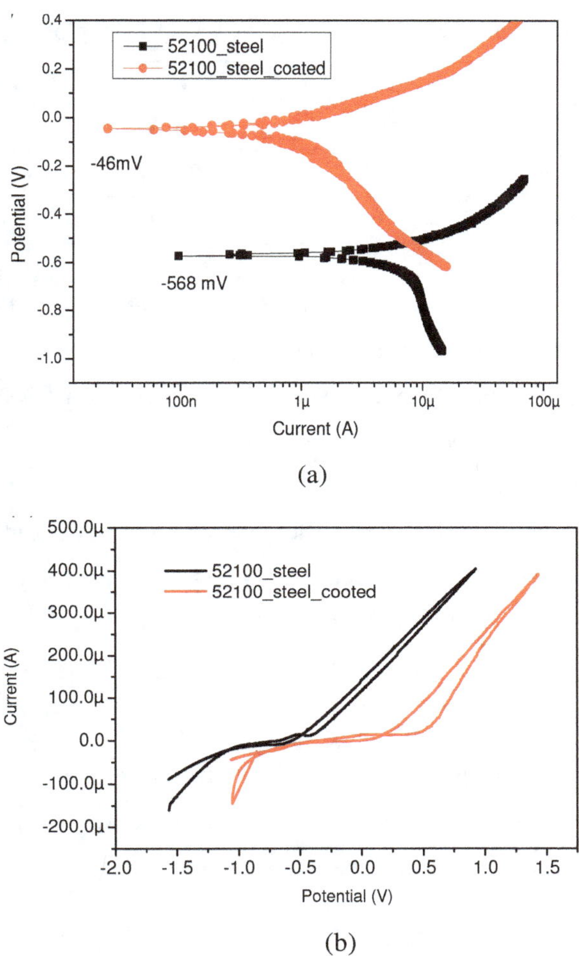

Figure 7.1 (a) Tafel extrapolation and (b) Cyclic polarization

More complete information about the electrochemical processes occurring at the electrodes or in the solution can be obtained using the cyclic polarization curve. In figure 7.1. (b), cyclic polarization diagrams are present. The anodic curves hardly overlapped the cathodic curves for both samples, characteristic of generalised corrosion, but at the same potential, the current density for coated 52100 steel is lower than for uncoated 52100 steel. It can be seen that the breakdown potential for the coated sample is around 0.6 V and for the uncoated sample around -0.4 V, higher electronegativity.

Table 7.1 Electrochemical parameters obtained from Tafel plots

Sample	E_0 (mV)	i_{corr} (mA)	Corrosion rate (mpy)
52100 uncoated	- 568	18.37	88.48
52100 coated	- 46	3.9	18.95

7.2 Determination of electrochemical corrosion behavior by electrochemical impedance spectroscopy

Figure 7.2 shows the Bode and Nquist plots obtained in HNO_3-H_2SO_4 acid rain solution. The fit parameters are specified in Table 7.2. The recorded impedance data agree with the model shown in Figure 7.2, Rs corresponds to the uncompensated resistance between the reference electrode and the working electrode, the passive surface capacitance CPE and Rp correspond to the passive surface resistance. The CPE impedance is defined by Eq:

$$Z_{CPE} = \frac{1}{Q(j\omega)^n}$$

(4.1)

where Q-is a constant proportional to the active area; ($<Q>$ = Ω-1 sn/cm2 $\equiv$ S.sn/cm2), ω is the angular frequency (ω = $2\pi f$, f- the frequency of the applied alternating current), j- is the imaginary number, j = $(-1)^{1/2}$ [83].

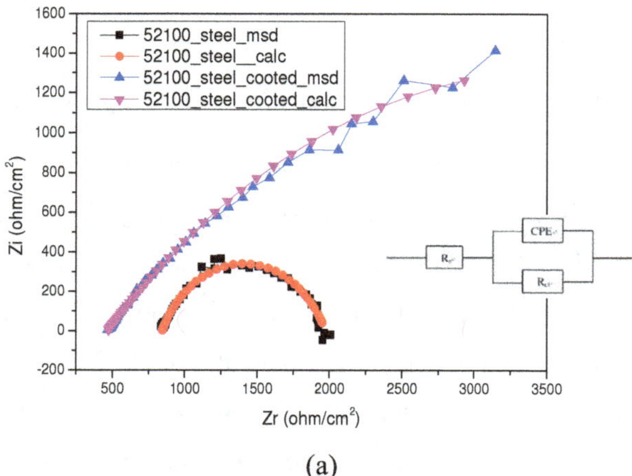

(a)

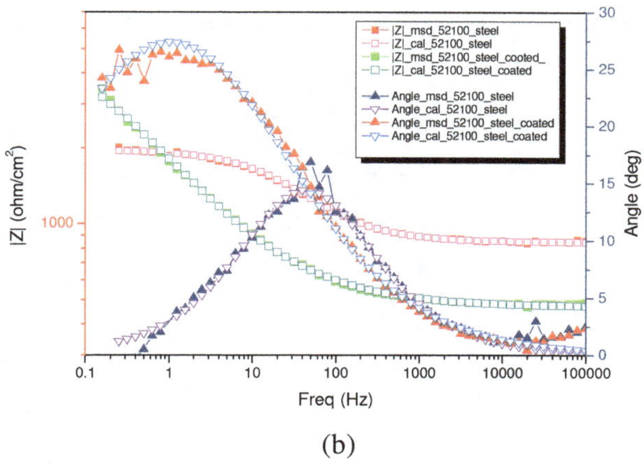

(b)

Figure 7.2 Nquist (a) and Bode (b) plots for coated and uncoated 52100 steel

Table 7.2 Simulated EIS Equivalent Circuit Fit Parameters

Sample / Parameter	R_s (Ω cm^2)	CPE (S.s^n/cm^2)	n	R_{ct} (Ω cm^2)
52100 uncoated	843	0.0000276	0.8	1130
52100 coated	469	0.000233	0.5034	6301

Capacity data obtained from EIS can be used to determine the compact passive film, which is related to the resistance and capacity of the passive film. The obtained value for Rct increases from 1130 (Ω cm^2) to 6301 Ω cm^2 for a coated 52100 sample. This can probably be attributed to the protection of the deposited layer. The Bode plot shows the existence of a single maximum on the curve (the phase angle as a function of the frequency of the applied signal), implying the existence of a single relaxation time constant. In general, three frequency regions that refer to high, medium and low values can be distinguished from the impedance spectrum. The high-frequency range of impedances at frequencies above 103 Hz highlights the values of solution resistance (Rsol). In the Bode plot, the impedance at medium frequencies represents the response of the surface layer, while at the lower frequency limit, the process information is related to the response of the substrate at the electrolyte interface.

The high impedance values (of the order of 106 Ω cm^2) obtained from medium and low frequencies for all coated samples suggest a high corrosion resistance in the electrolyte solution. The values obtained for the maximum phase angle open three frequency ranges (0.1-100 Hz), indicating a capacitive response for all samples and suggesting the presence of a surface layer on the surface acting as a barrier.

Spectrochemical analysis. Base material (AISI 52100 alloy steel)

At the level of the metal material, it can be seen that the surface is corroded, showing chemical corrosion compounds that seem to be of a different nature compared to steel, being estimated to be probably oxides and carbonates. The appearance of the compounds is the result of the interaction between the metallic material and the electrolyte solution. Corrosion is created over the entire surface of the sample as a generalization of it, and through several investigations its nature can be established. Figure 7.3 shows SEM images of the surface after open circuit potential (OCP), linear and cyclic potentiometry (LP and CP) and electro-impedance spectroscopy (EIS).

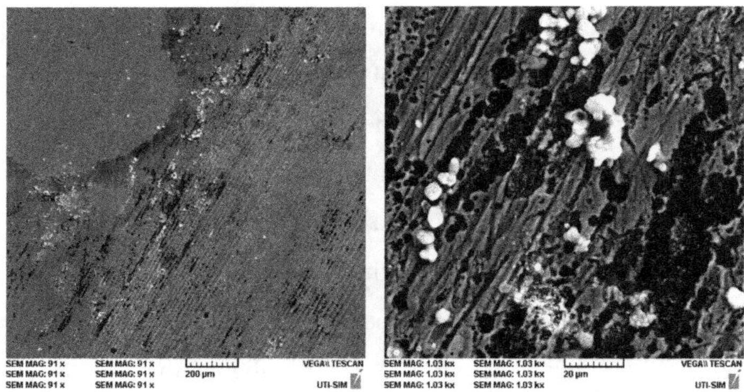

Figure 7.3 SEM images of the surface by open circuit potential (OCP), linear and cyclic potentiometry (LP and CP) and electro-impedance spectroscopy (EIS)

Through energy dispersive spectroscopy, in the chemical composition, in addition to the main elements, at the level of the sample surface, elements such as: oxygen resulting from the formation of oxides on the surface, nitrogen and sulfur, which are elements existing in the acid rain solution and produce compounds at surface. Table 7.3 shows the chemical composition of the surface after the electrochemical tests.

Table 7.3 Chemical composition of the surface after electrochemical tests

Element	At. No.	Netto	Mass [%]	Mass Norm [%]	Atom [%]	abs. error [%] (1sigma)	rel. error [%] (1sigma)
Iron	26	136520	94.03	92.96	81.88	2.41	2.57
Oxygen	8	1056	2.55	2.52	7.75	0.81	31.86
Nitrogen	7	374	2.41	2.38	8.38	0.87	35.92
Chromium	24	3704	1.50	1.48	1.40	0.08	5.341
Manganese	25	1188	0.66	0.65	0.58	0.07	10.79
	Sum		101.15	100	100		

At the surface level, the amount of oxygen is not very high, which concludes that the oxidation reactions are below the values of the reduction reactions at the surface, in this type of electrolyte. Small mass percentages are shown for both nitrogen and sulfur, based on the nitrides and sulfides that form during electro-corrosion processes. As it can be seen in Figure 7.4, the energies of the specified elements are given in the spectrum, highlighting two energies for Fe and Mn.

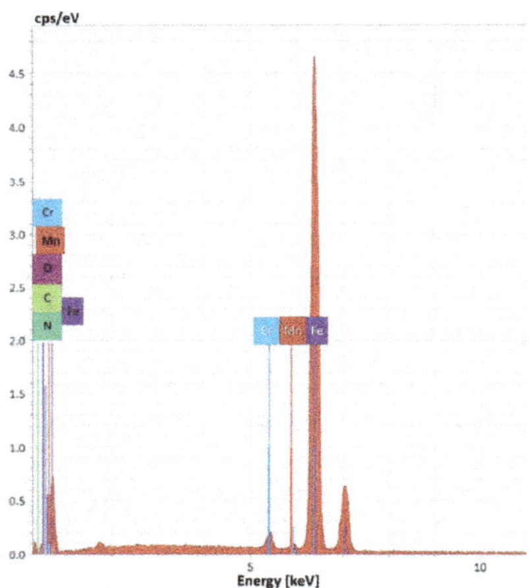

Figure 7.4 The energy spectrum of the elements identified on the corroded surface following the EIS test

On the surface of the sample, the generalised nature of the corrosion, confirmed by the elemental distribution on the elements, highlights the presence of oxides on the surface (the use of ultrasonic cleaning leads to their establishment). Figure 7.5 shows the elemental distribution on the corroded surface.

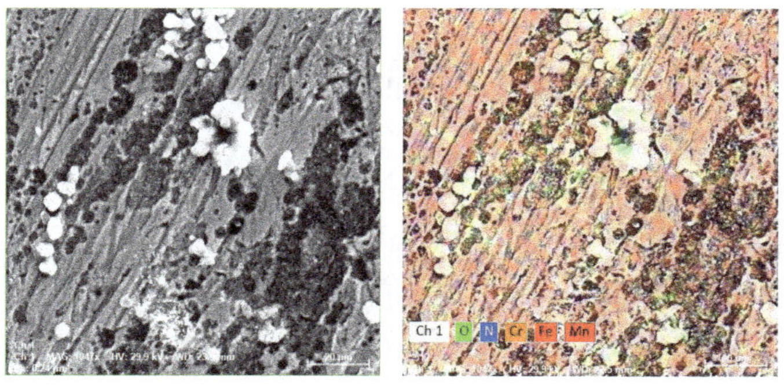

Figure 7.5 Elemental distribution on the corroded surface (all elements)

Spectrochemical analysis Material coated with Ni-Cr/C (WIP-C1) powders

Through micro-scale electro-corrosion tests, it is observed that the surface of the superficial layer deposited with WIP-C1 powders is intact. From the electrolyte solution, only a few compounds are present on the deposited surface and are the result of the interaction between the solution and the metal-ceramic system. Figure 7.6 shows SEM images of the surface, after open circuit potential (OCP), linear and cyclic potentiometry (LP and CP) and electro-impedance spectroscopy (EIS).

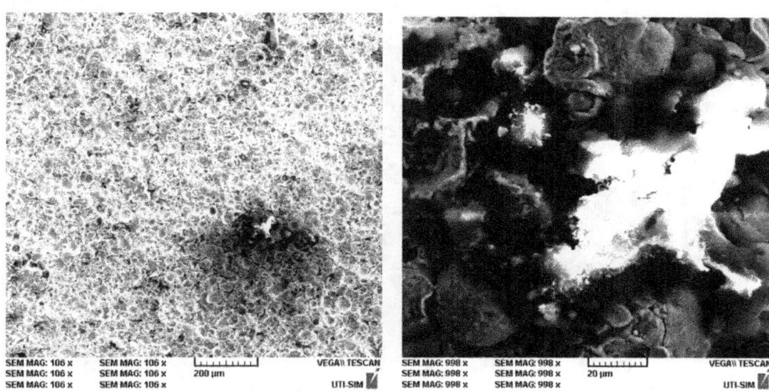

Figure 7.6 SEM images of the surface by open circuit potential (OCP), linear and cyclic potentiometry (LP and CP) and electro-impedance spectroscopy (EIS)

On the sample surface, at the micro-scale, few compounds can be observed. By analysing the chemical composition, in addition to the elements that make up the deposited layer (Cr, Ni, Si and C), elements in small quantities such as nitrogen, oxygen and sulfur are identified, as it can be seen in Figure 7.7. In contrast to the material of the base of the tested sample, it can be appreciated that the deposited superficial layer shows much less oxidation. Table 7.4 shows the chemical composition of the surface, after electrochemical tests.

Table 7.4 Chemical composition of the surface after electrochemical tests

Element	At. No.	Netto	Mass [%]	Mass Norm [%]	Atom [%]	abs. error [%] (1sigma)	rel. error [%] (1sigma)
Nickel	28	84670	80.83	69.62	56.99	2.05	2.53
Chromium	24	49486	27.16	23.39	21.62	0.75	2.76
Nitrogen	7	602	4.68	4.03	13.84	1.36	29.05
Oxygen	8	729	2.25	1.94	5.83	1.12	49.55
Silicon	14	1060	1.11	0.96	1.64	0.11	9.94
Sulfur	16	68	0.04	0.04	0.06	0.02	64.32
	Sum		116.09	100	100		

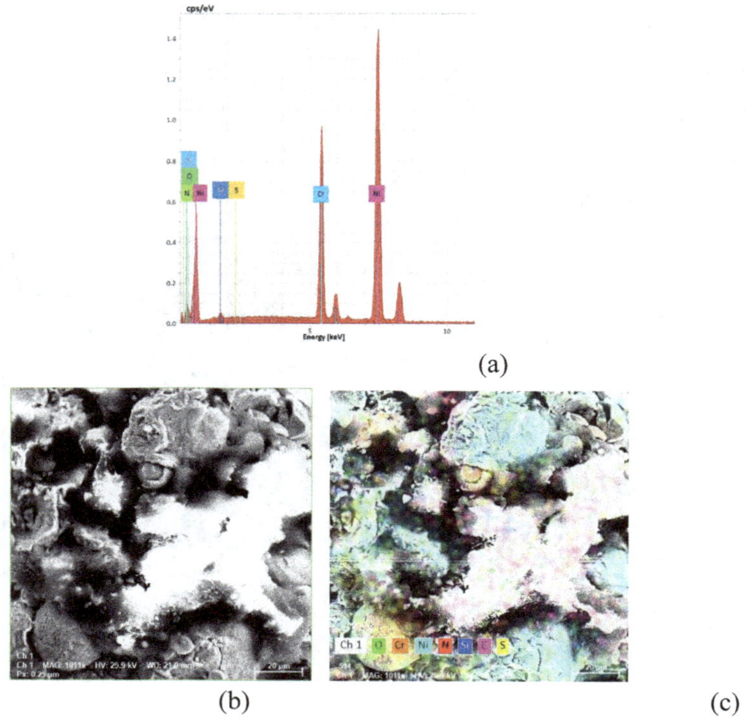

(a)

(b)   (c)

Figure 7.7 The energy spectrum of the elements identified on the corroded surface following the EIS test (a) and the elemental distribution on the corroded surface (all elements) (b) and (c)

7.3 Partial conclusions

1. The results obtained on the corrosion resistance are very good, these being determined for the coated samples, both by the linear potentiometry method and by the EIS tests.

2. The passive, stable, non-porous and non-cracked character of the coating was demonstrated by obtaining low values of the corrosion rate and a high resistance of the protection of the superficial deposited layer. This is also confirmed by the surface condition, after the corrosion tests, where it is observed that the coating surface remains intact.

CHAPTER 8 Final conclusions

1. Research shows that Ni/CrC forms a stable coating on 52100 alloy steel when deposited by the cold spray technique. Microstructural analysis shows that the coatings are dense and closely connected to the surface of the substrate, and through the simultaneous collision and deformation between the particles and the underlying substrate, the particles are deeply embedded, achieving this strong connection through a mechanical interlocking phenomenon.

2. The fatigue tests, through macroscopic and microscopic analysis, resulted in the following: highlighting the long-term fatigue propagation zone and the sudden crack propagation zone followed by sample rupture; the crack propagation zone is characterised by a fine structure specific to the sudden damage determined by the coarser particles; it was found that there are no significant pores or cracks at the level between the covering material and the base material; the cracks have as an initial stage, the surface of the base material, followed by the covering material and the interior of the base material, respectively; in the crack initiation phase, no detachments of the covering material due to it were observed; the deposited material behaved very well, over the entire crack propagation area, resulting in no detachment of it; Detachments of the deposited material can be observed at the level of the sudden propagation of cracks, there being gaps and detachments that are justified by the high values of the deformations created by the large difference in the specific deformations of the base material and the covering material. Sample no. 8 managed to exceed the reference value, so the test was stopped at a stress value of 438 MPa and a no. of 5451948 cycles, without observing any damage to the deposited material or its detachment, but it should be noted that the condition of the surfaces between the base material and the coating, where the first deformations appear at the level of the base material, can not be appreciated.

3. The hardness of the material increased from 28 HRC for the base material to 45 HRC, for the deposited layer.

4. Microscratch testing showed that the Ni/CrC layer has good microscratch resistance, as the friction coefficient showed high variability along the sliding path, with average values greater than 0.2. From the SEM images, it can be seen that the width of the scratch track is about 260 µm, the scratches are only on the coating material and the base material is not affected. It can be seen that the blade only removes the large asperities of the coating, which confirms the achievement of good resistance to microscratches.

5. The apparent coefficient of friction of the coatings was determined, both by linear and rotational motion. A short transition period was observed at the beginning of both tests, as the apparent contact area increased until it reached a steady state and flattened the large peaks of the coating asperities. Free particles increase the coefficient of friction, explaining the COF values between 0.4 and 0.7. The SEM images show that the tips of the asperities were removed and that the particles remaining on the sliding track acted as an abrasive material, leading to high values of the friction coefficient.

6. A significant improvement in corrosion resistance was obtained, as determined for coated samples by both linear potentiometry and EIS tests. The passive, stable, non-porous and non-cracked character of the coating was demonstrated by obtaining low values of the corrosion rate and by a high strength of the protection of the superficial layer deposited. This is also confirmed by the surface condition after the corrosion tests, where it is observed that the coating surface remains intact.

NAME OF SYMBOLS AND ABBREVIATIONS

RPG – Rocket Propelled Grenade;

HATE – High anti-tank explosive;

ARL – US Army Research Laboratory;

AMCA – Advanced Modular Composite Armor;

BH – Brinell Hardness;

UHH – Ultra High Hardness;

FEA – Finite Element Analysis;

RH – Rockwell Hardness;

CGS – Cold Gas Spraying;

APPJC – Atmospheric Pressure Plasma Jet Coating;

HVOF – High Velocity Oxygen Fuel;

PSVA – Plasma spraying in a Vacuum Atmosphere;

HVOF – High Velocity Oxygen Fuel;

WS – Warm Spray;

LPAPS – Low Pressure Atmosphere Plasma Spray;

D-GUN – the Detonation Gun Method;

CS – Cold Spray (Cold Spraying);

FS – Frictional Surface;

ASTM – American Society for Testing and Materials;

SAE – Society of Automobile Engineers;

AISI – American Iron and Steel Institute;

GNS – German National Standard;

ITS – Information Technology Standard;

BS – British Standard;

ES – European Standard;

RCF – Rolling Contact Fatigue;

SEM – Scanning Electron Microscope;

MSE – Microscope with Secondary Electrons;

HEEM – High Energy Electron Microscope;

OP-S – Oxide Polishing Suspension;

HPT – High Pressure Torsion;

TEM – Transmission Electron Microscopy;

PECVD – Plasma Enhanced Chemical Vapor Deposition;

FM – Frictional Mixing;

VH – Vickers Hardness;

HRTEM - High Resolution Transmission Electron Microscopy;

XRDA – X-ray Diffraction Analysis;

RC – Rapid Corrosion;

STVE – Scanning Technique with Vibrating Electrodes;

EIS – Electro-Impedance Spectroscopy;

SCE – Saturated Calomel Electrode;

HEA – High Entropy Alloys;

ABS – Acrylonitrile Butadiene Styrene;

PC – Polycarbonate;

PE – Polyethylene;

PP – Polypropylene;

UHMWP – Ultra High Molecular Weight Polyethylene;

FIB – Focused Ion Beam;

EDXRS – Energy Dispersive X-ray Spectroscopy;

ESEM – Environmental Scanning Electron Microscope;

HiVac – High Vacuum;

LowVac – Low Vacuum;

DM – Digital Microscope;

ITM – Inverted Trinocular Microscope;

RP – Polarization resistance;

EDS – Energy Dispersive Spectroscopy;

CAF – Coefficient of Apparent Friction;

D – Displacement;

FSC – Fast Surface Capacity;

OCP – Open Circuit Potential;

LP – Linear Potentiometry;

CP – Cyclic Potentiometry;

BIBLIOGRAPHY

[1]   George, G.M.T. Blindajul - principala armă a vehiculului de luptă Romania military [accesat la data 05.12.2021]

[2]   Wikipedia; (http://en.wikipedia.org/wiki/Vehicle_armour) [accesat la data 16.11.2021]

[3] Cristea, S. Contribuții la studiul comportării unor materiale de blindaj, la impactul cu proiectilul. Universitatea „Lucian Blaga" din Sibiu **2008**

[4]   https://ro.wikipedia.org/wiki/Blindaj [Accesat la data de 05.07.2022]

[5]   https://www.armyacademy.ro/reviste/3_2000/art14.html [Accesat la data de 05.07.2022]

[6]   https://ro.wikipedia.org/wiki/Bradley_Fighting_Vehicle [Accesat la data de 05.07.2022]

[7] Pintilie, D. Simularea comportării materialelor prin studii de dinamica explicită. *Cerc științific IMST* **2016**

[8]   https://ro.wikipedia.org/wiki/Blindaj_Chobham [Accesat la data de 05.07.2022]

[9]   http://www.military-today.com/tanks/t64.htm [Accesat la data de 05.07.2022]

[10] https://en.wikipedia.org/wiki/Fairchild_Republic_A-10_Thunderbolt_II [Accesat la data de 05.07.2022]

[11]   https://en.wikipedia.org/wiki/Mil_Mi-24 [Accesat la data de 05.07.2022]

[12]   https://www.britannica.com/technology/bomb-weapon [Accesat la data de 05.07.2022].

[13]   https://en.wikipedia.org/wiki/M1_Abrams [Accesat la data de 05.07.2022]

[14] Dwight, D.S.; William, A.G.; Mattew, S.B.; Stockman, R.; Koch, R. Balistic Testing of SSAB Ultra – High – Hardness Stell for Armour Applications. *Army Research Laboratory* **2008**

[15]   https://www.ssab.com/en [Accesat la data de 05.07.2022]

[16]   Locot. col. ing. Lungu, V. Blindajul Protecție și vulnerabilitate. Editura *Militară* **1980**

[17]   https://www.olysteel.com/products/alloy/mil-dtl-46100-mil-a-46100-armor-steel [Accesat la data de 05.07.2022]

[18] Poplawski, A.; Kedzierski, P.; Morka, A. Identification of Armox 500T steel failure properties in the modeling of perforation problems *Materials and Design* **2020, 190**

[19] Cristea, S. Contribuții la studiul comportării unor materiale de blindaj, la impactul cu proiectilul, *Universitatea „Lucian Blaga" din Sibiu, Facultatea de inginerie „Hermann Oberth" Catedra Știința și Ingineria Materialelor* **2008,** Teza de doctorat.

[20]   http://www.eurosteel-ro.com/iso/capvii-viii.pdf [Accesat la data de 26.07.2022]

[21] https://www.creeaza.com/referate/fizica/Recoacerea-tratamentul-termic-691.php [Accesat la data de 26.07.2022]

[22] Krugljakow, A.A.; Rogachev, S.O.; Lebedeva, N.V.; Sokolov, P.Y.; Arsenkin, A.M.; Khatkevich. On the nature of hot work hardening phenomenon in die steel with regulated austenitic transformation during exploitation. *Materials Science and Engineering:A* **2022**, **833**

[23] Davis, J.R.; Davis & Associates. Handbook of Thermal Spray Technology. *ASM International, All Rights Reserved. Handbook of Thermal Spray Technology* 2004

[24] Mitchell, R.D. Thermal Spray Coatings, Global turbine Sales Support. *Sulzer Metco (US) Inc., Westbury, New York* **2012**

[25] Harvinde, S. Experimental investigation of WC-12Co cold spray: Substrate hardness, bonding mechanism, powder type. *Materialstoday: proceedings* **2023**

[26] The Science and Engineering of Thermal Spray Coatings. John Wiley & Sons Ltd., Chichester, England. *Pawlowski, L. (Ed.)* **2008**

[27] Espallargas, N. Future Development of Thermal Spray Coatings. *1st Edition, Types, Designs, Manufacture and Applications* 2015

[28] Steenkiste, T.; Smith, J.R. Evaluation of coatings produced via kinetic and cold spraying processes. *J. Therm. Spray Technol.* **2004**, 13, 274–282

[29] Allkimov, A.P.; Papyrin, A.N; Kosarev, V.F.; Nesterovich, N.I.; Shushpanov, M.M. Gas–dynamic spray method for applying a coating. *US Patent 5 302 414, April 12.* **1994**

[30] Allkimov, A.P.; Papyrin, A.N.; Kosarev, V.F.; Nesterovich, N.I.; Shushpanov, M.M. Method and device for coating. *European Patent 0 484 533 B1, January 25.* **1995**

[31] Hermanek, F.J.; Thermal Spray Terminology and Company Origins. *ASM International, Materials Park, OH, USA.* **2001**

[32] Dykhuizen, R.C.; Smith, M.F. Gas dynamic principles of cold spray. *J. Therm. Spray Technol.* **1998**, 7 (2), 205–212

[33] Champagne, V.K. The Cold Spray Materials Deposition Process. *Woodhead Publishing Ltd, Cambridge, England* **2007**

[34] Stoltenhoff, T.; Kreye, H; Richter, H.J. An analysis of the cold spray process and its coatings. *J. Therm. Spray Technol.* **1994** 11 (4), 542–550

[35] Yamada, M.; Isago, H.; Nakano, H.; Fukumoto, M. Cold spraying of TiO_2 photocatalyst coating with nitrogen process gas. *J. Therm. Spray Technol.* **2010**. 19 (6), 1218–1223

[36] Yeom, H.; Sridharan, K. Cold spray technology in nuclear energy applications: A review of recent advances. *Ann. Nucl. Energy* **2021**, *150*, 107835

[37]   Neu, R.; Maier, H.; Boswirth, B.; Elgeti, S.; Greuner, H.; Hunger, K.; Kondas, J.; Muller, A. Investigations on cold spray tungsten/tantalum coatings for plasma facing applications. *Nucl. Mater. Energy* **2022**, *34*, 101343

[38]   Jan, C.; Monika, V.; Frantisek, L.; Martin, K.; Jan, K.; Reeti, R.S. Cold Sprayed Tungsten Armour for Tokamak First Wall. *Coatings* **2019**, *9*, 836

[39]   Agiwal, H.; Yeom, H.; Pocquette, N.; Sridharan, K.; Pfefferkorn, F.E. Friction surfacing and cold spray deposition for surface crack repair in austenitic stainless steels. *Mater. Today Comun.* **2022**, *33*, 104692

[40]   Yeom. H.; Dabney, T.; Pocquette, N.; Ross, K.; Pfefferkorn, F.E.; Sridharan, K. Cold spray deposition of 304L stainless steel to mitigate chloride-induced stress corrosion cracking in canisters for used nuclear fuel storage. *J. Nucl. Mater.* **2020**, *538*, 152254

[41]   Wan, W.; Li, W.; Wu, D.; Qi, Z.; Zhang, Z. New insights into the effects of powder injector inner diameter and overhang length on particle accelerating behaviour in cold spray additive manufacturing by numerical simulation. *Surf. Coat. Technol.* **2022**, *444*, 128670

[42]   Lupoi, R.; Neill, W. Powder stream characteristics in cold spray nozzles. *Surf. Coat. Technol.* **2011**, *206*, 1069–1076

[43]   Singh, H.; Sidhu, T.S.; Kalsi, S.B.S. Cold spray technology: Future of coating deposition processes. *Frat. Integrità Strutt.* **2012**, *22*, 69–84

[44]   VRC Metal Systems. 600 N Ellsworth Rd, Box Elder, SD 57719, United States. (https://vrcmetalsystems.com/)

[45]   Espalargas, N. Introduction to thermal spray coatings. *Future Development of Thermal Spray Coatings.* **2015**

[46]   Fauchais, P. Current status and future directions of thermal spray coatings and techniques. *Future Development of Thermal Spray Coatings* **2015**

[47]   Fauchais, P.; Heberlein, J.; Boulos, M.; Thermal Spray Fundamentals. Springer NY, USA. **2014**, 1600 pages

[48]   Grasset, G. Projection thermique et codeposition electrolytique, deux alternatives. *in 1st International Meeting on Thermal Spraying, L. Powlowski (Ed.) ENSCL, Lille, 3-4 December* **2003**, pp. 197 – 202 unpublished conference proceedings

[49]   www.sciencereviews2000.co.uk [Accesat la data de 15.08.2022]

[50]   Fagoaga, I.; Barykin, G.; Juan. J.; Soroa, T.; Vaquero, C. The high frequency pulse detonation (HFPD) spray process. *C., Coatings, A., Inasmet, F., 1999. In: Thermal Spray 1999: United Thermal Spray Conference (DVS-ASM).* **1999** pp. 282 – 287

[51]   Carlson, R.; Heberlein, J. Effects of operating parameters on high definition single wire arc spraying. *in Thermal Spray 2001: New Surfaces for a New Millenium, C.C Berndt, K.A. Khor and E. Lugscheider (Eds), ASM International, Materials Park, OH, USA.* **2001**, pp. 447–453

[52]   Beardsley, M.B.; Happoldt, P.G.; Kelley, K.C. Thermal Barrier Coatings for Low Emission, High Efficiency Diesel Engine Applications- *SAE International 1999-01-2255. April 26-28,* **1999**

[53]   Vaidya, A.; Srinivasan, V.; Streibl, T.; Friis, M.; Chi, W.; Sampath, S.; Tucker Jr. C. *(EdASM Handbook. Thermal Spray Technology, vol. 5A. 2008. Process maps for plasma spraying of yttria-stabilised zirconia: an integrated approach to design, optimisation and reliability. Mater. Sci. Eng.* **2013**, A 497, 239–253

[54] Muehlberger, E. Industrial plasma processing technology. *In: Proc. 1st Plasma Technik Symp., vol. 3. Plasma Technik, Wohlen.* **1988**, pp. 105–118

[55]   Ambuhl, P.; Meyer, P. Thermal coating technology in controlled atmospheres (Cham-ProTM). *In: Lugscheider, E., Kammer, P.A. (Eds.). Proceedings of the ITSC. DVS, Dusseldorf, Germany.* **1999**, pp. 291–292

[56] Freslon, A. Plasma spraying at controlled temperature and atmosphere. *In: Berndt, C.C., Sampath, S. (Eds.), In: Thermal Spray: Science and Technology. ASM International, OH, USA.* **1995** pp. 57–63

[57]   http://www.hvaf.com/ [Accesat la data de 21.08.2022]

[58]   Goutier, S.; Vardelle, M.; Labbe, J.C.; Fauchais, P. Flattening and cooling of millimeterand micrometer-sized alumina drops. *J. Therm. Spray Technol.* **2011**, 20 (1–2), 59–67

[59]   Bandyopadhyay, R.; Nyle´n, P. A computational fluid dynamic analysis of gas and particle flow in flame spraying. *J. Therm. Spray Technol.* **2003**, 12 (4), 492–503

[60]   Thorpe, M.L.; Richter, H.J. A pragmatic analysis and comparison of HVOF processes. *J. Therm. Spray Technol.* **1992**, 1 (2), 161–170

[61]   Sobolev, V.V.; Fagoaga, I. Warm spray: A new promising technology of the coating deposition. *San Sebastian/E.* **2002**

[62]   Seiji, K.; Jin, K.; Makoto, W, Hiroshi, K. A novel coating process based on high-velocity impact of solid particles. *Warm spraying.* **2008**

[63]   https://continentalsteel.com/carbon-steel/grades/alloy-52100/ [Accesat la data de 29.03.2022]

[64]   http://m.ro.lksteelpipe.com/52100-bearing-steel [Accesat la data de 29.03.2022]

[65]   Zhang, R.Y.; Muftu, S. Elastic impact of spherical particle with a long, stationary, fixed Timoshenko beam. J. Sound Vib. 2021, 495, 115892

[66]   Budde, L.; Biester, K.; Lammers, M.; Hermsdorf, J.; Kaierle, S.L. Overmeyer. Influence of process parameters on single weld seam geometry and process stability in Laser Hot-Wire Cladding of AISI 52100. Adv. Ind. Manuf. Eng. 2023, 7, 1001122

[67]   ASTM A295; Standard Specification for High-Carbon Anti-Friction Bearing Steel. ASTM: West Conshohocken, PA, USA, 2017

[68]   DIN 17230; Ball and Roller Bearing Steels. Publisher: Deutsches Institut fur Normung E.V, 1980

[69]   IT G4805; High Carbon Chromium Bearing Steel. Publisher: Japanese Standards Association, 2023

[70]   BS 970; Wrought Steels in the Form of Bars, Billets and Forgings Up to 6 in. Ruling Section for Automobile and General Engi-neering Purpose. Publisher: British Standards Institution, 1955

[71]   Panda, A.; Sahoo, A.K.; Kumar, R.; Das, R.K. A review on machinability aspects for AISI 52100 bearing steel *Materialstoday: Proceedings.* **2020**, **23** (3) pp 617-621

[72]   Smelova, V.; Schwedt, A.; Wang, L.; Holweger, W.; Mayer, J. Electron microscopy investigations of microstructural alterations due to classical Rolling Contact Fatigue (RCF) in martensitic AISI 52100 bearing steel *International Journal of Fatigue.* **2017**, 98, *pp 142-154*

[73]   Areitioaurtena, M.; Segurajauregi, U.; Fisk, M.; Cabello, M.J.; Ukar, E. Influence of induction hardening residual stresses on rolling contact fatigue lifetime *International Journal of Fatigue,* **2022**, *159*

[74]   Duan, C.; Qu, S.; Hu, X.; Jia, S.; Li, X. Evaluation of the influencing factors of combined surface modification on the rolling contact fatigue performance and crack growth of AISI 52100 steel *Wear,* **2022**, 494 – 495

[75]   Kiranbabu, S.; Tung, P.Y.; Sreekala, L.; Prithiv, T.S.; Hickel, T.; Pippan, R.; Morsdorf, L.; Herbig, M. Cementite decomposition in 100Cr6 bearing steel during high-pressure torsion: Influence of precipitate composition, size, morphology and matrix hardness *Materials Science and Engineering:A.* **2022**, 833

[76]   Qin, Y.; Mayweg, D.; Tung, P.Y.; Pippan, R.; Herbig. Mechanism of cementite decomposition in 100Cr6 bearing steels during high pressure torsion *Acta Materialia.* **2020**, 201, pp 79-93

[77]   Solis, R.J.; Rodrigues, M. A.; Colis, C.J.; Neville, A. Parametric optimisation of friction and wear of a multi-layered a-C:H coating on AISI52100 steel *Materials Letters,* **2022**, 318

[78] Rahbar, K,A.; Abdollah, Z. A.; Hadavi, M.M.; Benerji, A.; Alpas, A.; Gerlich, A.P. Effects of friction stir processing on wear properties of WC-12%Co sprayed on 52100 steel. *Mater. Des.* **2015**, *86*, 98–104

[79] https://order.powdersondemand.com/ [Accesat la data de 31.08.2022]

[80] Young, B.; Heelan, J.; Langan, S.; Siopis, M.; Walde, C.; Birt, A. Novel characterisation techniques for additive manufacturing powder feedstock *Metals,* **2021**, 11, 1-19

[81] https://www.oerlikon.com/metco/en/products-services/ [Accesat la data de 01.09.2022]

[82] Wong, W.; Rezaeian, A.; Irissou, E.; Legoux, J.G.; Yue, S. Cold spray characteristics of commercially pure Ti and Ti-6Al-4V *Adv.Mater.Res.* **2010**, 89-91 pp 639-644

[83] Herrera, J.E.J.; Bousser, E.; Schmitt, T.; Klemberg, S.J.E.; Martinu, L. Effect of plasma interface treatment on the microstructure, residual stress profile, and mechanical properties of PVD TiN coatings on Ti-6Al-4V substrates *Surface and Coatings Technology.* **2021**, 413

[84] Jun, Y.L.; Ayan, B.; Adrian, W.Y.T.; Wen, S.Xu.S.; Wei, Z.; Pio, J.B.; Feng, L.; Rjia, L.; Teng, M.L.; Chris, B.B. Understanding the microstructural evolution of cold sprayed Ti-6Al-4V coatings on Ti-6Al-4V substrates *Applied Surface Science,* **2018**, 459, pp 492-504

[85] Lioma, D.; Sacks, N.; Botef, I.; Cold gas dynamic spraying of WC – Ni cemented carbide coating. *International Journal of Refractory Metals and Hard Materials*, March **2015**

[86] Melendez, N.M.; Narulkar, V.V.; Fisher, G.A.; McDonald, A.G.; Effect of reinforcing particles on the wear rate of low - pressure cold - sprayed WC - based MMC coatings, *Wear* 30 August **2013**

[87] Silva, F.C.; Cinca, N.; Dosta, S.; Cano, I.G.; Couto, M.; Guilemany, J.M.; Benedetti, A.V. Corrosion behaviour of WC - Co coatings deposited by cold gas spray onto AA 7075 - T6, Corrosion Science, 15 May **2018**

[88] Couto, M.; Dosta, S.; Torrell, M.; Fernandez, J.; Guilemany, J.M. Cold spray deposition of WC-17 and 12Co cermets onto aluminium. *Surf. Coat. Technol.* **2013**, *235*, 54–61

[89] Ashokkumar, M.; Thirumalaikumarasamy, D.; Thirumal, P.; Barathiraja. Influences of Mechanical, Corrosion, erosion and tribological performance of cold sprayed Coatings A review *Materials Today: Proceedings.* **2021**, 46, 17 pp 7581-7587

[90] Longqiang, Z.; Bopin, X.; Mengxuan, Z.; Shunjie, H.; Guodong, Z.; Fabrication and characterisation of Al-Zn-Cu composite coatings with different Cu contents by cold spraying *Surface Engineering.* **2020**, 36, pp 1090-1096

[91]   Zhang, Z.; Liu, F.; Han, E.H.; Xu, L. Mechanical and corrosion properties in 3.5% NaCl solution of cold sprayed Al-based coatings. *Surf. Coat. Technol.* **2020**, *385*, 12537

[92]   Ma, J.; Wen, J.; Li, Q.; Zhang, Q. Electrochemical polarisation and corrosion behaviour of Al-Zn-In based alloy in acidity and alkalinity solutions. *Int. J. Hydrog. Energy* **2013**, *38*, 14896–14902

[93]   Mohammed, A.A., *A* newly synthesised glycine derivative to control uniform and pitting corrosion processes of Al induced by SCN-anions-Chemical, electrochemical and morphological studies. *Corros. Sci.* **2010**, *52*, 3243–3257

[94]   Kumar, S.; Kumar, M.; Jindal, N. Overview of cold spray coating application comparisons: a critical review *World Journal of Engineering,* **2020**, **17** 27-51

[95]   Pathak, S.; Saha, G . Development of Sustainable Cold Spray Coatings and 3D Additive Manufacturing Components for Repair/Manufacturing Applications: A Critical Review *Coatings.* **2017, 7** 122

[96]   Gadow, R.; Killinger, A.; Stiegler, N. Hidroxyapatite coatings for biomedical applications deposited by different thermal spray techniques *Surface and Coatings Technology.* **2010**, 205, 1157-64

[97]   Zou, Y.; Qiu, Z.; Huang, C.; Zeng, D.; Lupoi, R.; Zhang, N.; Yin, S. Microstructure and tribological properties of Al2O3 reinforced FeCoNiCrMn high entropy alloy composite coatings by cold spray. *Surf. Coat. Technol.* **2022**, *434*, 128205

[98]   Deng. N.; Qu, D.; Zhang, K.; Liu, G.; Li, S.; Zhou, Z. Simulation and experimental study on cold sprayed W-Cu composite with high retainability of W using core-shell powder. Surface and Coatings Technology. **2023**, 466, 129639

[99]   https://jteg.ncms.org/wp-content/uploads/2019/09/02-Aaron-JTEG-CS-4-28-2020.pdf [Accesat la data de 03.09.2022]

[100]   https://phillipscorp.com/federal/2020/06/24/vrc-army-phillips-federal/ [Accesat la data de 03.09.2022]

[101]   Che, H.; Liberati, A.C.; Chu, X.; Chein, M.; Nobari, A.; Vo, P.; Yue, S. Metallisation of polymers by cold spraying with low melting point powders. *Surf. Coat. Technol.* **2021**, *418*, 127229

[102] Che, H.; Phuong, V.; Stephen, Y. Investigation of cold spray on polymers by single particle impact experiments. *J. Therm. Spray Technol.* **2019**, 28, 135–143

[103] Che, H.; Chu, X.; Vo, P.; Yue, S. Metallisation of various polymers by cold spray. *J. Therm. Spray Technol.* **2018**, 27, 169–178

[104] Chen, M.; Che, H.; Yue, S. Exploring surface preparation for cold spraying on polymers. *Surf. Coat. Technol.* **2022**, *450*, 128993

[105] Northeastern University, 360 Huntington Ave, Boston, Massachusetts, MA 02115, United States (https://www.northeastern.edu/)

[106] Marius Bibu, Metode şi tehnici de analiză structurală a materialelor metalice, Ed. Editura Universităţii ''Lucian Blaga'' Sibiu, **2000**, pg. 196

[107] https://www.thermofisher.com/ro/en/home/electron-microscopy/products/scanning-electron-microscopes.html?cid=cmp-05676-c1d3&utm_source=google-ads&utm_medium=cpc&utm_campaign=ms_xmarket_gl_sem_search_google-adwords_2022_05&utm_term=%7Bkeyword%7D&gad=1&gclid=Cj0KCQjwiIOmBhDjARIsAP6YhSXJw9Drdipq7epchMtXQMFb4WOd5IoiZMP9lPRJ3u6HH5x35-68IYkaAvz4EALw_wcB [Accesat la data de 26.07.2023].

[108] https://mec.tuiasi.ro/despre/departamentul-de-inginerie-mecanica-mecatronica-si-robotica/centrul-de-organe-de-masini-si-mecatronica/laboratorul-de-studiul-materialelor [Accesat la data de 26.07.2023]

[109] Munteanu Corneliu, Istrate Bogdan, Lupescu Stefan-Constantin. Ştiinţa şi ingineria materialelor: Îndrumar de laborator (222 pagini), Ed. Europlus. Galaţi, **2019**, ISBN 978-606-628-207-9

[110] Corneliu Munteanu, Mihai Stefan, Constantin Baciu, Nicanor Cimpoesu - METODE DIFRACTOMETRICE ŞI MICROSCOPIE OPTICĂ ŞI ELECTRONICĂ ÎN STUDIUL MATERIALELOR (263 pagini), Editura Tehnopress Iaşi 2008, ISBN 978-973-702-563-0

[111] http://clay.uga.edu/courses/8550/XRD.html [Accesat la data de 27.07.2023]

[112] http://www.ecomet.pub.ro/infrastructura/expertize-mecano-metalurgice/microscop-electronic-cu-baleiaj-tip-quanta-inspect-f-producator-fei-philips-olanda/ [Accesat la data de 30.07.2023]

[113] Corneliu Munteanu - STUDIUL MATERIALELOR – Structură - Metode de investigare - Echilibru termo - dinamic în sistemele materiale- Solidificarea materialelor metalice, Editura „Gh. ASACHI" Iaşi – 2001, 302 pag., tiraj 200 exemplare, ISBN 973-8050-92-8

[114] Munteanu C., Studiul Materialelor, Ed. Gh.Asachi, Iaşi, 2010.

[115] Hopulele I., Cimpoesu N., Nejneru C., Metode de analiză a materialelor – microscopie şi analiză termică, Editura Tehnopress (acreditată CNCSIS), Iaşi, 2009 ISBN 978-973-702-673-6

[116] http://clay.uga.edu/courses/8550/XRD.html [Accesat la data de 01.08.2023]

[117] https://www.malvernpanalytical.com/en/products/category/x-ray-diffractometers?utm_source=google&utm_medium=cpc&utm_campaign=EN%20-%20Panalytical%20-%20Product%20Category%20-%20X-ray%20Diffractometers&utm_term=%2Bpanalytical%20%2Bdiffractometer&utm_content=54063894460&gad=1&gclid=Cj0KCQjw2qKmBhCfARIsAFy8buK0YPJ0Q3LrW3xZsC2oxqArJ5TVCTLzfU5rVX_LehL9qalxLiOB2-saAoD7EALw_wcB [Accesat la data de 05.08.2023]

[118] https://mec.tuiasi.ro/despre/departamentul-de-inginerie-mecanica-mecatronica-si-robotica /centrul-de-organe-de-masini-si-mecatronica/laboratorul-de-tribologie/ [Accesat la data de 05.08.2023]

[119] https://www.instron.com/en/search-results?ss360Query=8801[Accesat la data de 06.08.2023]

[120] https://mec.tuiasi.ro/diverse/V.Goanta/2_Obos_Laborator_2017.pdf[Accesat la data de 06.08.2023].

[121] Benea Lidia, Laborator: Electrochimie şi Coroziune. Universitatea Dunărea de Jos, din Galaţi

[122] Mardare Laurenţiu. Teza de doctorat: Studiul degradării prin coroziune şi imbunatăţirea metodelor de protecţii anticorozive a structurilor metalice în mediul marin, **2019,** Galaţi

[123] https://pdf.directindustry.com/pdf/hach-lange/voltalab-catalog/5842-126431.html [Accesat la data de 20.09.2023]

[124] Goanta, V.; Munteanu, C.; Muftu, S.; Istrate, B.; Schwartz, P.; Boese, S.; Ferguson, G.; Moraras, C.I. Evaluation of the Fatigue Behaviour and Failure Mechanisms of 52100 Steel Coated with WIP-C1 (Ni/CrC) by Cold Spray. Materials, 2022, 15(10), 3609

[125] Gateman, S.M.; Gharbi, O.; de Melo, H.G.; Ngo, K.; Turmine, M.; Vivier, V. On the use of a constant phase element (CPE) in electrochemistry. Curr. Opin. Electrochem. 2022, 36, 101133